Algebra 2 Made Easy Handbook

Common Core Standards Edition

W9-BGO-417

By:
Mary Ann Casey

B. S. Mathematics, M. S. Education

Acknowledgments

My sincere thanks to those who helped me put *Algebra II Made Easy, Common Core Edition*, together. The people who contributed include Kimberly Knisell, Director of Math and Science in the Hyde Park, NY School District for her organizing expertise and to Jennifer Criser-Eighmy, Director of Humanities in Hyde Park, for proofreading the grammar and punctuation; to my daughter, Debra Rainha who teaches math at Andover (MA) High School and to her colleague, Stephanie Ragucci who teaches AP Statistics at Andover High. Her assistance was invaluable. Marin Malgieri contributed to the unit on Probability and Statistics, and additional proofreading was done by Laura Denkins. Graphics designer, Julieen Kane, at Topical Review Book Company always does a great job getting the graphs and diagrams done correctly as well as putting my drafts into the necessary form to be published. Keith Williams, owner of Topical Review Book Company, is always easy to work with and very patient. I am honored to have my work published by "the "Little Green Book" company – familiar to many for providing Regents Examination study materials since 1936.

Introduction

This quick reference guide, or "how to do it book" It is not meant as a curriculum guide to the Common Core Standards, nor is it meant to be used as a textbook. The Common Core Standards (CCS) involve new methods of teaching and learning and use various methods for solving mathematical problems. Implementing new teaching techniques and providing deeper understanding of the content of the course is the job of the classroom teacher. As the Algebra II CCS are implemented, it is my hope that this student friendly book will assist students in becoming "college and career ready", as this is one of the main goals of our educational process.

Sincerely,

MaryAnn Casey,
B.S. Mathematics, M.S. Education

ALGEBRA 2 MADE EASY
Common Core Standards Edition
Table of Contents

ALGEBRA 2 MADE EASY
Common Core Standards Edition

Table of Contents

Unit 1

POLYNOMIALS, RATIONAL, AND RADICAL RELATIONSHIPS

- Perform arithmetic operations with complex numbers.

- Use complex numbers in polynomial identities and equations.

- Interpret the structure of expressions.

- Understand the relationship between zeros and factors of polynomials.

- Use polynomial identities to solve problems.

- Rewrite rational expressions.

- Understand solving equations as a process and explain the reasoning.

- Solve equations and inequalities in one variable.

- Solve systems of equations.

- Analyze functions using different representations.

- Translate between the geometric description and the equation for a conic section.

- Extend the properties of exponents to rational exponents.

REAL NUMBERS AND EXPONENTS

In past math courses, we have worked with the real numbers. The real numbers consist of all the numbers on the number line. In this course, we will also study numbers in the set of complex numbers.

The two main subsets of the "reals" are the rational numbers and the irrational numbers. The rational numbers include all those numbers that can be written in the form of a fraction where both the numerator and the denominator are integers. The irrational numbers include the square roots of imperfect squares, decimal numbers that do not repeat and do not terminate, Pi (π), and e which is the base of natural logarithms. (See page 143.)

A complex number can be written in the form $a + bi$ where a and b are real numbers, i is the imaginary unit and $i = \sqrt{-1}$.
(See Chapter 1.3 – Imaginary and Complex Numbers)

EXPRESSIONS WITH EXPONENTS

Note: All the rules that apply to numbers with exponents are also used for expressions containing variables.

Bases and Exponents: An exponent indicates how many times to use its base as a factor. The base of the exponent is the number, variable, or parenthesis directly to the left of the exponent. If the expression to the left of the exponent is a parenthesis, the exponent is applied to the entire content of the parenthesis. Evaluating with an exponent is often referred to as "raising" a number to a power.

Review of Properties of Exponents:

$$a^x \bullet a^y = a^{x+y}$$

$$\left(a^x\right)^y = a^{xy}$$

$$\frac{a^x}{a^y} = a^{x-y}, \ a \neq 0$$

$$\left(ab\right)^x = a^x b^x$$

$$\left(\frac{a}{b}\right)^x = \frac{a^x}{b^x}, \ b \neq 0$$

$$a^0 = 1, \ a \neq 0$$

$$a^{-1} = \frac{1}{a}, \ a \neq 0$$

1.1

Examples **Bases and Exponents**

❶ $3^4 = 3 \cdot 3 \cdot 3 \cdot 3 = 81$

❷ $-3^4 = -(3 \cdot 3 \cdot 3 \cdot 3) = -81$ —— The base is 3. The negative sign is not directly to the left of the exponent, therefore it is not affected by the exponent.

❸ $(-3)^4 = (-3)(-3)(-3)(-3) = 81$ —— Here the base is the contents of the parenthesis, -3.

❹ $5x^2 = 5 \cdot x \cdot x = 5x^2$ —— Only the x is the base for the exponent, 2.

❺ $(5x)^2 = (5x)(5x) = 25x^2$ —— $(5x)$ is directly to the left of the exponent, 2, so the entire parenthesis is the base and is multiplied by itself to evaluate the expression.

The same rule applies if the base is a fraction or a polynomial.

Example

$$\left(\frac{2}{3}\right)^2 = \left(\frac{2}{3}\right) \cdot \left(\frac{2}{3}\right) = \frac{4}{9} \text{ is different than } \frac{2^2}{3} = \frac{2 \cdot 2}{3} = \frac{4}{3}$$

$$x(x+2)^2 = x(x+2)(x+2) = x^3 + 4x^2 + 4x, \quad (x+2) \text{ is the base.}$$

Negative Exponents: A negative exponent directs us to use the reciprocal of the base raised to the indicated exponent. The negative sign has *nothing to do with the positive or negative value* of the base. Once we translate the base into its reciprocal, the negative sign on the exponent disappears. The positive exponent is then applied as usual.

Examples

❶ $2^{-3} = \left(\frac{1}{2}\right)^3 = \left(\frac{1}{2}\right)\left(\frac{1}{2}\right)\left(\frac{1}{2}\right) = \frac{1}{8}$

❷ $3(x)^{-2} = 3\left(\frac{1}{x}\right)^2 = 3\left(\frac{1}{x}\right)\left(\frac{1}{x}\right) = \frac{3}{x^2}$

❸ $\left(\frac{x}{2}\right)^{-2} = \left(\frac{2}{x}\right)^2 = \left(\frac{2}{x}\right)\left(\frac{2}{x}\right) = \frac{4}{x^2}$

❹ $(-5)^{-2} = \left(-\frac{1}{5}\right)^2 = \left(-\frac{1}{5}\right)\left(-\frac{1}{5}\right) = \frac{1}{25}$

Fractional Exponents: A fractional exponent directs us to take a specific root of a base, and then use the root as a base to be raised to a power. The denominator of the fraction indicates the root needed, and the numerator tells to what power to raise the root (base). (This process can be done in either order, but finding the root first, then raising it to a power involves numbers that are usually easier to handle.)
Reminder: The "root" of a radical sign is called the index.
(See Unit 1.2 – Radicals)

Examples Fractional Exponents

❶ $8^{\frac{2}{3}} = \left(\sqrt[3]{8}\right)^2 = 2^2 = 4$

❷ $\left(\frac{25}{9}\right)^{\frac{3}{2}} = \left(\sqrt{\frac{25}{9}}\right)^3 = \left(\frac{5}{3}\right)^3 = \frac{125}{27}$

The radical sign without an index indicates the principal square root.

Negative Fractional Exponents: The process for using negative exponents is combined here with the process for evaluating expressions with fractional exponents. Take care of the negative part of the exponent first by changing the original base to its reciprocal. Then find the root indicated by the denominator and raise it to the power of the numerator.

Examples

❶ $(64)^{-\frac{2}{3}} = \left(\frac{1}{64}\right)^{\frac{2}{3}} = \left(\sqrt[3]{\frac{1}{64}}\right)^2 = \left(\frac{1}{4}\right)^2 = \frac{1}{16}$

❷ $(-8)^{-\frac{5}{3}} = \left(\frac{1}{-8}\right)^{\frac{5}{3}} = \left(\sqrt[3]{\frac{1}{-8}}\right)^5 = \left(\frac{1}{-2}\right)^5 = \frac{1}{-32} = -\frac{1}{32}$

❸ $(x)^{-\frac{3}{2}} = \left(\frac{1}{x}\right)^{\frac{3}{2}} = \left(\sqrt{\frac{1}{x}}\right)^3$ or $\frac{1}{\sqrt{x^3}}$. When simplified and

rationalized, $\frac{1}{\sqrt{x^3}} = \frac{1}{x\sqrt{x}} \cdot \frac{\sqrt{x}}{\sqrt{x}} = \frac{\sqrt{x}}{x^2}$.

1.1

MULTIPLYING AND DIVIDING EXPRESSIONS WITH LIKE BASES

- To **multiply** expressions with like bases, keep the base and add the exponents.

Examples

❶ $5^3 \cdot 5^4 = 5^7$ — To evaluate, multiply 5 times itself 7 times.
$5 \cdot 5 \cdot 5 \cdot 5 \cdot 5 \cdot 5 \cdot 5 = 78,125$

❷ $6^{\frac{2}{3}} \cdot 6^{\frac{1}{2}} = 6^{\left(\frac{2}{3}+\frac{1}{2}\right)} = 6^{\left(\frac{4}{6}+\frac{3}{6}\right)} = 6^{\frac{7}{6}}$ — In order to add these fractional exponents, a common denominator is required. A calculator would be used to evaluate the expression which is approximately equal to 8.088.

❸ $3^4 \cdot 5^2$ — These expressions do not have the same base, therefore the rule of adding exponents cannot be used. To evaluate, process each number separately, then multiply.
$3^4 = 81;\ 5^2 = 25;\ 25 \cdot 81 = 2025$

- To **divide** expressions with like bases, keep the base and subtract the exponents.

Examples

❶ $\dfrac{7^5}{7^3} = 7^{(5-3)} = 7^2$ — To evaluate, multiply $7 \cdot 7 = 49$

❷ $\dfrac{10^2}{2^3} = \dfrac{100}{8} = \dfrac{25}{2}$ — The bases are not alike, so the rule of subtracting the exponents cannot be used. Process each number first, then simplify.

❸ $\dfrac{(5x)^3}{(2x)^2} = \dfrac{5^3 \cdot x^3}{2^2 \cdot x^2} = \dfrac{125}{4}\left(x^{(3-2)}\right) = \dfrac{125}{4}x \ or \ \dfrac{125x}{4}$ — The 5 and 2 are not like bases, but the variable x is in both expressions. Process them separately since the rules of exponents for division can be used with x, but not with 5 and 2.

Note: See additional examples from Unit 1.3 on page 12.

Use of the radical symbol with an index and a radicand requires finding the root of the radicand that is indicated by the index. If no index is shown, it is automatically a 2 and indicates a square root. The radical symbol itself stands for the principal root of the radicand which is the positive root of an even root. (If there is a negative radicand with an odd index, the principal root would be negative.) If the index is even, a negative radicand has no real root. If there is a negative sign in front of the radical, the sign of the root is changed. The number to the left of the radical is the coefficient. Do the work on the radical, then multiply by the coefficient.

Examples

❶ $5\sqrt[3]{81} = 5\left(\sqrt[3]{27 \bullet 3}\right) = 5\left(3\sqrt[3]{3}\right) = 15\sqrt[3]{3}$ — The radical is left in the problem to indicate that the cube root of 3 is not a rational number.

❷ $\sqrt[3]{-8} = -2$ — Negative radicand, odd index – principal root is negative.

❸ $\sqrt{-64}$ — Negative radicand, even index – no real root.

❹ $-\sqrt{16} = -4$ — The coefficient here is negative which makes the answer negative.

❺ $-\sqrt[3]{-64} = -(-4) = 4$ — The principal cube root, (index = 3) of –64 is –4, but the coefficient of –1 makes the answer positive.

SIMPLIFYING RADICALS

Simplified Radical Form: A radical in simplest form has no number or variable left under the radical that is a power of the indicated root. (See example 1 below.) If it is a fraction, it has no radical in the denominator. The following examples are simplified radicals:

Examples ❶ $\sqrt{20} = 2\sqrt{5}$ ❸ $\sqrt{8x^3} = 2x\sqrt{2x}$

❷ $\sqrt{16x^2} = 4x$ ❹ $\dfrac{5}{\sqrt{6}} \bullet \dfrac{\sqrt{6}}{\sqrt{6}} = \dfrac{5\sqrt{6}}{6}$

Irrational Real Numbers: A square root radical that has been simplified and still has a number under the radical is irrational. Examples 3 and 4 are irrational numbers. If a number is the sum or difference of an integer and an irrational number, it is considered irrational.

Example $3 + \sqrt{2}$

Algebra 2 Made Easy – Common Core Edition

Steps — How to simplify a radical:

1) Determine all the prime factors of the radicand.

2) Based on the root indicated by the index, called the "*nth* root", make groups of *n* prime factors. Take the *nth* root of each group and place it outside the symbol. Repeat for all factors until no groups remain.

3) Multiply the factors outside the radical together and multiply the factors inside the radical (the radicand) back together.

Examples

❶ $\sqrt[3]{54x^3} = \sqrt[3]{3 \cdot 3 \cdot 3 \cdot 2 \cdot x \cdot x \cdot x} = 3x\sqrt[3]{2}$

The cube root of $3 \cdot 3 \cdot 3 = 3$, the cube root of $xxx = x$.

❷ $6\sqrt{75x^5} = 6\sqrt{5 \cdot 5 \cdot 3 \cdot x \cdot x \cdot x \cdot x \cdot x} = 6(5)(x^2)\sqrt{3x} = 30x^2\sqrt{3x}$

Simplifying Fraction—Radical Combinations: Reduce the fraction if possible. Canceling can be done between two coefficients or between two radicands. A coefficient cannot be canceled with a radicand.

• Separate the fraction into two radicals – one containing the numerator and the other containing the denominator. Then simplify each radical as shown above. Cancel the coefficients or radicals if possible. If a radical remains in the denominator, the fraction must be "rationalized" by multiplying the numerator and the denominator by that radical.

Examples ❶ $\sqrt{\dfrac{16}{18}} = \sqrt{\dfrac{8}{9}} = \dfrac{\sqrt{8}}{\sqrt{9}} = \dfrac{2\sqrt{2}}{3}$

❷ $\sqrt{\dfrac{14}{28}} = \sqrt{\dfrac{1}{2}} = \dfrac{\sqrt{1}}{\sqrt{2}} \cdot \dfrac{\sqrt{2}}{\sqrt{2}} = \dfrac{\sqrt{2}}{2}$

❸ $\sqrt{\dfrac{24x^3}{36x^4}} = \sqrt{\dfrac{2}{3x}} = \dfrac{\sqrt{2}}{\sqrt{3x}} \cdot \dfrac{\sqrt{3x}}{\sqrt{3x}} = \dfrac{\sqrt{6x}}{3x}$

If a fraction already contains a radical in the numerator and another in the denominator, they can be combined into one radical if it makes the problem easier to work with.

Example $\dfrac{\sqrt{16x^2}}{\sqrt{8x^4}} = \sqrt{\dfrac{16x^2}{8x^4}} = \sqrt{\dfrac{2}{x^2}} = \dfrac{\sqrt{2}}{x}$

See Rationalizing the Denominator on page 10.

OPERATIONS WITH RADICALS

Adding and Subtracting: Only radicals that have the same index and the same radicand can be added or subtracted. (They are called "like" radicals, similar to "like terms".) The coefficients are added or subtracted and the index, radical sign, and radicand stay unchanged. It is often possible to simplify a radical in order to make the radicands alike.

Steps:

1) Compare the indexes on each radical. If they are not all the same, the radicals cannot be combined with others.
2) Compare radicands. If they are not the same, try to simplify each one so they are alike.
3) Add/Subtract the coefficients of "like" radicals. The radicand remains unchanged.
4) If several different radicands are involved, (or if the indexes are not the same) the sum or difference may have two or more terms in the answer.

Examples

❶ $3\sqrt{5} + 3\sqrt{20} = 3\sqrt{5} + 3\left(2\sqrt{5}\right) = 3\sqrt{5} + 6\sqrt{5} = 9\sqrt{5}$ — In this example, $3\sqrt{20}$ must be simplified before it can be added to $3\sqrt{5}$.

❷ $2\sqrt{7} + 3\sqrt{14} = 2\sqrt{7} + 3\sqrt{14}$ — Since neither radical can be simplified the answer has two terms.

❸ $\sqrt{6} - \sqrt[3]{6} = \sqrt{6} - \sqrt[3]{6}$ — The radicals do not have the same index, so they cannot be subtracted. This answer also stays in two parts.

❹ $\sqrt{50} - 6\sqrt{2} + \sqrt{5} = 5\sqrt{2} - 6\sqrt{2} + \sqrt{5} = -\sqrt{2} + \sqrt{5}$ — Notice that $\sqrt{50}$ can be simplified and combined with $6\sqrt{2}$, but $\sqrt{5}$ has a different radicand and cannot be combined with the others.

Multiplying and Dividing: The radicands do not have to be equal when multiplying or dividing.

Steps:

1) Check to make sure the indexes are alike. If they are not, change to fractional exponents and use the laws of exponents to multiply or divide.
2) Multiply or divide the coefficients.
3) Multiply or divide the radicands.
4) Simplify the results. Make sure to rationalize any denominator that contains a radical.

Algebra 2 Made Easy – Common Core Edition

1.2

Examples Multiplying:

❶ $3\sqrt{5} \cdot 7\sqrt{10} = 21\sqrt{50} = 21\left(5\sqrt{2}\right) = 105\sqrt{2}$ — After multiplying the radicands, simplify the radical. Then multiply the coefficients, 5 and 21, together.

❷ $5\left(3\sqrt{7}\right) = 15\sqrt{7}$ — 5 and 3 are both coefficients and can be multiplied.

* When the radicands are alike but the indexes are not the same, change the form of the each expression into a base with a fractional exponent. Then keep the base as it is and use the laws of exponents -- find a common denominator for the exponents and add the exponents. Simplify.

❶ $\sqrt[3]{15} \cdot \sqrt[2]{15} = \left(15^{\frac{1}{3}}\right)\left(15^{\frac{1}{2}}\right) = 15^{\left(\frac{1}{3}+\frac{1}{2}\right)} = 15^{\left(\frac{2}{6}+\frac{3}{6}\right)} = 15^{\frac{5}{6}} = \sqrt[6]{15^5} \; or \; \left(\sqrt[6]{15}\right)^5$

❷ $\sqrt[3]{(7x)^2} \cdot \sqrt{7x} = (7x)^{\frac{2}{3}}(7x)^{\frac{1}{2}} = (7x)^{\frac{2}{3}+\frac{1}{2}} = (7x)^{\frac{4}{6}+\frac{3}{6}} = (7x)^{\frac{7}{6}}$

Examples Dividing:

❶ $\dfrac{6\sqrt{10}}{2\sqrt{5}} = \dfrac{6}{2}\sqrt{\dfrac{10}{5}} = 3\sqrt{2}$

❷ $5\sqrt{\dfrac{36}{16}} = \dfrac{5\sqrt{36}}{\sqrt{16}} = \dfrac{5(6)}{4} = \dfrac{30}{4} = \dfrac{15}{2} \; or \; 7.5$ — Separate the numerator and denominator into two radicals and simplify each. Multiply the coefficient, 5, by the simplified numerator. Divide, or leave as a reduced fraction.

❸ $\dfrac{3x^2\sqrt{8y}}{x\sqrt{2y}} = \dfrac{3x^2}{x}\sqrt{\dfrac{8y}{2y}} = 3x\sqrt{4} = 3x(2) = 6x$

* When the bases are the same but the indexes are not, change each expression to fractional exponent form and subtract the exponents, keeping the base as it is. A common denominator for the exponents is needed to subtract the exponents.

Examples

❶ $\dfrac{\sqrt{5}}{\sqrt[3]{5}} = \dfrac{5^{\frac{1}{2}}}{5^{\frac{1}{3}}} = 5^{\left(\frac{1}{2}-\frac{1}{3}\right)} = 5^{\left(\frac{3}{6}-\frac{2}{6}\right)} = 5^{\frac{1}{6}} = \sqrt[6]{5}$

Radicals

MULTIPLYING POLYNOMIALS THAT CONTAIN RADICALS

Distribute each term in one polynomial to each term in the other. Follow the rules for multiplying radicals. (Use FOIL for multiplying binomials with radicals.) Simplify.

Examples

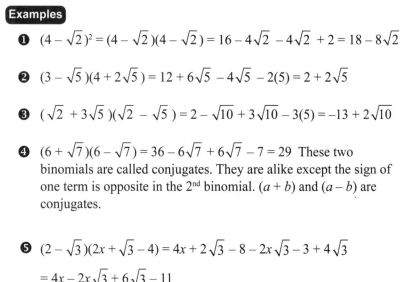

❶ $(4 - \sqrt{2})^2 = (4 - \sqrt{2})(4 - \sqrt{2}) = 16 - 4\sqrt{2} - 4\sqrt{2} + 2 = 18 - 8\sqrt{2}$

❷ $(3 - \sqrt{5})(4 + 2\sqrt{5}) = 12 + 6\sqrt{5} - 4\sqrt{5} - 2(5) = 2 + 2\sqrt{5}$

❸ $(\sqrt{2} + 3\sqrt{5})(\sqrt{2} - \sqrt{5}) = 2 - \sqrt{10} + 3\sqrt{10} - 3(5) = -13 + 2\sqrt{10}$

❹ $(6 + \sqrt{7})(6 - \sqrt{7}) = 36 - 6\sqrt{7} + 6\sqrt{7} - 7 = 29$ These two binomials are called conjugates. They are alike except the sign of one term is opposite in the 2nd binomial. $(a + b)$ and $(a - b)$ are conjugates.

❺ $(2 - \sqrt{3})(2x + \sqrt{3} - 4) = 4x + 2\sqrt{3} - 8 - 2x\sqrt{3} - 3 + 4\sqrt{3}$

$= 4x - 2x\sqrt{3} + 6\sqrt{3} - 11$

RATIONALIZING THE DENOMINATOR

A fraction in simplest form cannot have a radical in its denominator. To remove it, rationalize the denominator.

Rationalize a Denominator: Multiply the numerator and denominator by a factor that will remove the radical from the denominator. A single radical, or the conjugate of the denominator may be needed.

Conjugates: A pair of binomials where one binomial is the sum of two terms, the other is the difference of two terms.

Examples

❶ $(x + 3)$ and $(x - 3)$

❷ $(x - \sqrt{2})$ and $(x + \sqrt{2})$

❸ $(a + b)$ and $(a - b)$

- **Monomial Denominator:** Multiply the numerator and denominator by the radical in the denominator. Simplify.

Examples

❶ $\dfrac{3}{\sqrt{2}} \cdot \dfrac{\sqrt{2}}{\sqrt{2}} = \dfrac{3\sqrt{2}}{2}$

❷ $\dfrac{5}{3\sqrt{7}} \cdot \dfrac{\sqrt{7}}{\sqrt{7}} = \dfrac{5\sqrt{7}}{3(7)} = \dfrac{5\sqrt{7}}{21}$

❸ $\dfrac{2\sqrt{5}}{3\sqrt{2}} \cdot \dfrac{\sqrt{2}}{\sqrt{2}} = \dfrac{2\sqrt{10}}{6} = \dfrac{\sqrt{10}}{3}$

- **Binomial Denominator:** Multiply the numerator and denominator by the conjugate of the denominator. Use FOIL. Combine like terms; reduce if possible.

Examples

❶ $\dfrac{2}{5+\sqrt{3}} \cdot \left(\dfrac{5-\sqrt{3}}{5-\sqrt{3}}\right) = \dfrac{2\left(5-\sqrt{3}\right)}{25-3} = \dfrac{\overset{1}{\cancel{2}}\left(5-\sqrt{3}\right)}{\underset{11}{\cancel{22}}} = \dfrac{5-\sqrt{3}}{11}$

❷ $\dfrac{6+\sqrt{5}}{7-\sqrt{5}} \cdot \left(\dfrac{7+\sqrt{5}}{7+\sqrt{5}}\right) = \dfrac{\left(6+\sqrt{5}\right)\left(7+\sqrt{5}\right)}{49-5} = \dfrac{42+13\sqrt{5}+5}{44} = \dfrac{47+13\sqrt{5}}{44}$

❸ $\dfrac{3+\sqrt{2}}{5+\sqrt{7}} \cdot \left(\dfrac{5-\sqrt{7}}{5-\sqrt{7}}\right) = \dfrac{\left(3+\sqrt{2}\right)\left(5-\sqrt{7}\right)}{25-7} = \dfrac{15-3\sqrt{7}+5\sqrt{2}-\sqrt{14}}{18}$

❹ $\dfrac{2-2\sqrt{3}}{1-2\sqrt{3}} \cdot \left(\dfrac{1+2\sqrt{3}}{1+2\sqrt{3}}\right) = \dfrac{\left(2-2\sqrt{3}\right)\left(1+2\sqrt{3}\right)}{1-4(3)} = \dfrac{2+2\sqrt{3}-4(3)}{-11} =$

$$\dfrac{-10+2\sqrt{3}}{-11} \quad or \quad \dfrac{10-2\sqrt{3}}{11}$$

$$\boxed{1.2}$$

Examples Examples for Unit 1.1 and Unit 1.2.

① Given that $\left(\sqrt[5]{x}\right)^5 = x$ write the expression equivalent to $\sqrt[5]{x}$ as a variable with an exponent.

Solution: $\left(\sqrt[5]{x}\right)^5 = x^{\frac{5}{5}} = x$ which makes $\left(\sqrt[5]{x}\right) = x^{\frac{1}{5}}$

② Which of the following expressions is equal to n?

1. $(n^a)^{\frac{-1}{a}}$ 2. $(n^a)^a$ 3. $\dfrac{n^a}{n^a}$ 4. $(n^a)^{\frac{1}{a}}$

Solution: Translate each possible choice to an equivalent form to compare.

1. $(n^a)^{\frac{-1}{a}} = n^{\frac{-a}{a}} = n^{-1} = \dfrac{1}{n}$ 3. $\dfrac{n^a}{n^a} = n^{(a-a)} = n^0 = 1$

2. $(n^a)^a = n^{a^2}$ 4. $(n^a)^a = n^{\frac{a}{a}} = n^1 = n$

The correct choice is answer #4.

③ Simplify the expression $(27x^4y^9)^{\frac{1}{3}}$

Solution: Apply the exponent 1/3 to each of the terms inside the parenthesis. $(27x^4y^9)^{\frac{1}{3}} = \left(27^{\frac{1}{3}}\right)\left(x^{\frac{4}{3}}\right)\left(y^{\frac{9}{3}}\right) = 3xy^3\sqrt[3]{x}$

④ Simplify the following expression: $\dfrac{9\left(x^{\frac{1}{3}}y^{\frac{-2}{3}}\right)^4}{\sqrt{16x^2y^2}}$

Solution: Apply the exponent of 4 to each of the terms inside the parenthesis in the numerator. Simplify the denominator. Combine the numerator and denominator using the rules of exponents.

$$\dfrac{9\left(x^{\frac{1}{3}}y^{\frac{-2}{3}}\right)^4}{\sqrt{16x^2y^2}} = \dfrac{9\left(x^{\frac{4}{3}}y^{\frac{-8}{3}}\right)}{4xy} = \dfrac{9x^{\left(\frac{4}{3}-1\right)}y^{\left(\frac{-8}{3}-1\right)}}{4} = \dfrac{9x^{\frac{1}{3}}y^{\frac{-11}{3}}}{4} = \dfrac{9y\sqrt[3]{xy^2}}{4} = \dfrac{9\sqrt[3]{xy^2}}{4y^3}$$

⑤ Simplify the following expression: $64^{\frac{-5}{6}}$

Solution: The negative exponent means to use the reciprocal of the base. The fractional exponent indicates that the answer will be the 6th root of the base to the 5th power.

$$64^{\frac{5}{6}} = \left(\dfrac{1}{64}\right)^{\frac{5}{6}} = \sqrt[6]{\left(\dfrac{1}{64}\right)^5} = \left(\dfrac{1}{2}\right)^5 = \dfrac{1}{32}$$

Note: Other methods of solution are possible in examples 1-5 above.

Algebra 2 Made Easy – Common Core Edition

IMAGINARY AND COMPLEX NUMBERS

Imaginary and Complex Numbers: $i = \sqrt{-1}$, and it is called the imaginary unit. The square root of a negative number does not have a real number solution, and a simplified square root radical cannot contain a negative radicand. Use i to simplify a square root radical that contains a negative radicand. Always simplify by processing the $\sqrt{-1}$ before simplifying the other numbers. If the original radicand is a negative perfect square, the simplified answer will have only a number, n, and i. If it is irrational, in simplified form it will be $i\sqrt{n}$.

Examples

❶ $\sqrt{-3} = \sqrt{-1} \cdot \sqrt{3} = i\sqrt{3}$

❷ $\sqrt{-75} = \sqrt{-1} \cdot 5\sqrt{3} = 5i\sqrt{3}$

❸ $\sqrt{-64} = \sqrt{-1} \cdot \sqrt{64} = 8i$

❹ $\sqrt{-12} = \sqrt{-1} \cdot \sqrt{12} = i\sqrt{12} = 2i\sqrt{3}$

Powers of i: The powers of i have a cyclical rotation. When simplifying an imaginary number using i, we never use it with an exponent higher than 1 because all the other powers of i can be simplified to 1, -1, or $-i$.

In any problem containing i with an exponent, i must be simplified so it has no exponent other than 1.

$i^0 = 1$ Any number or variable $\neq 0$, raised to the zero power $= 1$.

$i^1 = i$ Remember that $i = \sqrt{-1}$ which is defined as i.

$i^2 = \left(\sqrt{-1}\right)^2 = -1$

$i^3 = i \cdot i^2 = -i$

$i^4 = i^2 \cdot i^2 = (-1)(-1) = 1$

$i^5 = i^4 \cdot i = i$

$i^6 = i^5 \cdot i = i^2 = -1$

$i^7 = i^6 \cdot i = -i$

Circle diagram: $i^0 = 1$, $i^1 = i$, $i^2 = -1$, $i^3 = -i$

This circle diagram shows the cyclical nature of the powers of i. Start at $i^0 = 1$ and read clockwise.

Note: The correct exponent for i is the remainder when the given exponent is divided by 4.

i^{35} : $35 \div 4 = 8$ remainder 3. i^{35} is equivalent to i^3 which equals $-i$

$i^{35} = i^{32} \cdot i^3 = (1)i^3 = -i$

Simplifying an Imaginary Number Using _i_: Process the exponents according to the operation indicated. The laws of exponents are used with the exponents of _i_.

Examples Simplify: $i^5 \cdot i^2$

Steps: 1) Add the exponents: $i^5 \cdot i^2 = i^{5+2} = i^7$

       2) Break i^7 down: $i^7 = i^4 \cdot i^3$

       3) Evaluate: $(1)(-i) = -i$

❶ $i^{23} = i^{20} \cdot i^3 = (1)i^3 = -i$ ❸ $5i^{17} = 5 \cdot i^{16} \cdot i = 5 \cdot i^0 \cdot i = 5(1)i = 5i$

❷ $\dfrac{i^{14}}{i^{12}} = i^{14-12} = i^2 = -1$ ❹ $-(i^3)^3 = -i^9 = -(i^0)(i^1) = -(1)(i) = -i$

COMPLEX NUMBERS

A number can be real, pure imaginary, or complex. A complex number contains two parts – a real number and an imaginary number. It is written in the form $a + bi$ where a and b are real numbers and $i = \sqrt{-1}$. A number consisting only of bi is considered "pure imaginary". (Some texts refer to all numbers as complex as they consider that a could be zero, or b could be zero but the number is still in $a + bi$ form since zero is a real number.)

A complex number is written in two parts. If it involves a fraction, it cannot be written as a single fraction. Writing $\dfrac{4 + 3i}{5}$ is not correct. Write it as $\dfrac{4}{5} + \dfrac{3}{5}i$, keeping the real number separate from the imaginary number.
• Location of _i_: Write it before a radical but after a rational number. $i\sqrt{3}$ or $5i$

Coefficient of _i_: If _i_ has a coefficient, the related radical has the same coefficient.

Examples Simplify and write in $a + bi$ form:

❶ $5\sqrt{-1} = 5i$

❷ $2\sqrt{-16} = 2(4i) = 8i$

❸ $17 + \sqrt{-8} = 17 + 2i\sqrt{2}$

❹ $10 - \sqrt{-5} + 12 - \sqrt{-500} - \sqrt{5} = 22 - i\sqrt{5} - 10i\sqrt{5} - \sqrt{5} = 22 - \sqrt{5} - 11i\sqrt{5}$

[$-11i\sqrt{5}$ and $-\sqrt{5}$ are not like terms and cannot be combined. The _i_ makes them unlike terms. The two real numbers are written first, then the imaginary number.]

Algebra 2 Made Easy – Common Core Edition

Examples

❶ Simplify the following expression. Justify each step using the commutative, associative, and distributive properties.

$$(8 - 5i)(2 + 5i)$$

Steps:1) Multiply binomials using the distributive property. $(8-5i)(2+5i) \Rightarrow 16+40i-10i-25i^2$

2) Associative property to collect like terms.

$16 + 30i - 25i^2$

3) Substitution of −1 for i.

$16 + 30i - 25(-1) \quad 16 + 30i + 25$

4) Commutative property followed by associative property.

$16 + 25 + 30i = 41 + 30i$

❷ What are the complex roots of the following polynomial?

$f(x) = (x^2 + 3)(x^2 - 5)(x^2 + 16)(3x + 6)$

Solution: A complex root is given in the form $a + bi$. The roots are found when $f(x) = 0$. Using the zero product property, each factor in the polynomial is equal to zero when $f(x) = 0$. Set each factor equal to zero and solve to find the roots.

$x^2 + 3 = 0$	$x^2 - 5 = 0$	$x^2 + 16 = 0$	$3x + 6 = 0$
$x^2 = -3$	$x^2 = 5$	$x^2 = -16$	$3x = -6$
$x^2 = \pm\sqrt{-3}$	$x = \pm\sqrt{5}$	$x = \pm\sqrt{-16}$	$x = -2$
$x = \pm i\sqrt{3}$		$x = \pm 4i$	

The complex roots for this polynomial are $x = \pm i\sqrt{3}$ and $x = \pm 4i$. In these roots, $a = 0$. The other roots, $x = \pm\sqrt{5}$ and $x = -2$ are real roots.

❸ **Simplify:** $i(2 - 3i) - (3 + 2i)^2$

Solution: $i(2 - 3i) - (3 + 2i)^2 = 2i - 3i^2 - [9 + 12i + 4i^2]$
$-10i - 7i^2 - 9 = -10i - 7(-1) - 9 = -2 - 10i$

❹ If the voltage produced by each of 2 batteries is $-5 + 2i$ and their combined voltage is increased three times, what is the total voltage produced?

Solution: $3[2(-5 + 2i)] = 6(-5 + 2i) = -30 + 12i$

OPERATIONS WITH COMPLEX NUMBERS

When adding, subtracting, multiplying and dividing complex numbers in the form $a + bi$, i is treated just like a variable. In the final answer, if i is raised to a power, it must be simplified.

Examples

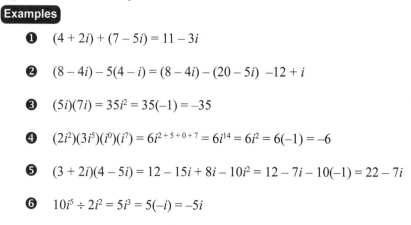

❶ $(4 + 2i) + (7 - 5i) = 11 - 3i$

❷ $(8 - 4i) - 5(4 - i) = (8 - 4i) - (20 - 5i)\ \ -12 + i$

❸ $(5i)(7i) = 35i^2 = 35(-1) = -35$

❹ $(2i^2)(3i^5)(i^0)(i^7) = 6i^{2+5+0+7} = 6i^{14} = 6i^2 = 6(-1) = -6$

❺ $(3 + 2i)(4 - 5i) = 12 - 15i + 8i - 10i^2 = 12 - 7i - 10(-1) = 22 - 7i$

❻ $10i^5 \div 2i^2 = 5i^3 = 5(-i) = -5i$

When dividing or simplifying a problem containing $a + bi$ it is necessary to rationalize the denominator if it contains i because $i = \sqrt{-1}$. If i is part of a monomial denominator, multiply the numerator and denominator by i. If it is part of a binomial denominator, multiply the numerator and denominator by the conjugate of $a + bi$ which is $a - bi$. Be sure to simplify and put the answer in $a + bi$ form when finished.

Examples

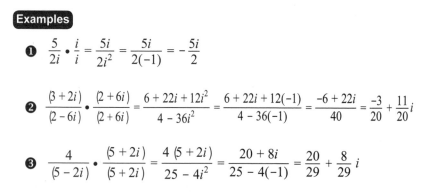

❶ $\dfrac{5}{2i} \cdot \dfrac{i}{i} = \dfrac{5i}{2i^2} = \dfrac{5i}{2(-1)} = -\dfrac{5i}{2}$

❷ $\dfrac{(3 + 2i)}{(2 - 6i)} \cdot \dfrac{(2 + 6i)}{(2 + 6i)} = \dfrac{6 + 22i + 12i^2}{4 - 36i^2} = \dfrac{6 + 22i + 12(-1)}{4 - 36(-1)} = \dfrac{-6 + 22i}{40} = \dfrac{-3}{20} + \dfrac{11}{20}i$

❸ $\dfrac{4}{(5 - 2i)} \cdot \dfrac{(5 + 2i)}{(5 + 2i)} = \dfrac{4(5 + 2i)}{25 - 4i^2} = \dfrac{20 + 8i}{25 - 4(-1)} = \dfrac{20}{29} + \dfrac{8}{29}i$

COMPLEX SOLUTIONS AND THE QUADRATIC FORMULA

- Solve using the **Quadratic Formula**: $x = \dfrac{-b \pm \sqrt{b^2 - 4ac}}{2a}$

  Steps:
  1) Identify a, b, and c.
  2) Substitute the numbers into the formula.
  3) "Do the math." Perform the indicated operations and simplify. Be careful with the signs.
  4) Leave in simplest radical form, $a + bi$ form, or as a decimal using the calculator if required.

Examples

❶ $3x^2 - 5x + 2 = 0$

$a = 3, \; b = (-5), \; c = 2$

$x = \dfrac{-(-5) \pm \sqrt{(-5)^2 - 4(3)(2)}}{2(3)}$

$x = \dfrac{5 \pm \sqrt{25 - 24}}{6}$

$x = \dfrac{5 + 1}{6} = 1, \; x = \dfrac{5 - 1}{6} = \dfrac{2}{3}$

Solution : $x = \{1, \frac{2}{3}\}$

❷ $4x^2 - 4x - 11 = 0$

$a = 4, \; b = (-4), \; c = (-11)$

$x = \dfrac{-(-4) \pm \sqrt{(-4)^2 - 4(4)(-11)}}{2(4)}$

$x = \dfrac{4 \pm \sqrt{16 + 176}}{8} = \dfrac{4 \pm \sqrt{192}}{8}$

$x = \dfrac{4 + 8\sqrt{3}}{8}, \quad x = \dfrac{4 - 8\sqrt{3}}{8}$

$x = \dfrac{1 + 2\sqrt{3}}{2}, \quad x = \dfrac{1 - 2\sqrt{3}}{2}$

Solution $= \left\{ \dfrac{1 + 2\sqrt{3}}{2}, \; \dfrac{1 - 2\sqrt{3}}{2} \right\}$

❸ $x^2 - 2x + 6 = 0$

$a = 1, \; b = (-2), \; c = 6$

$x = \dfrac{-(-2) \pm \sqrt{(-2)^2 - 4(1)(6)}}{2(1)}$

$x = \dfrac{2 \pm \sqrt{4 - 24}}{2}$

$x = \dfrac{2 \pm \sqrt{-20}}{2}$

$x = \dfrac{2 \pm i\sqrt{20}}{2} = \dfrac{2}{2} \pm \dfrac{2i\sqrt{5}}{2}$

$x = 1 + i\sqrt{5}, \quad x = 1 - i\sqrt{5}$

$SS = \left\{ 1 + i\sqrt{5}, \; 1 - i\sqrt{5} \right\}$

Examples

❶ What are the solutions for $(x - 3)^2 = -16$?

Solution: $(x - 3)^2 = -16$

$$x - 3 = \pm\sqrt{-16}$$

$$x = 3 \pm 4i$$

❷ A graph of the function $f(x) = 4x^2 + 20$ has no x-intercepts. Susan says that $f(x)$ has no solution. Her team member, Bob, says that it has 2 solutions. Determine who is correct by finding algebraically if $f(x)$ has any solutions or not, and what the solutions are if they exist.

Solution: Set $f(x)$ equal to zero and solve for x.

$4x^2 + 20 = 0$

$4(x^2 + 5) = 0$

$x^2 = -5$

$x = \pm\sqrt{-5}$

$x = i\sqrt{5}$ and $x = -i\sqrt{5}$

Conclusion: $f(x)$ does have 2 roots so Bob is correct. The roots are imaginary roots, not real roots. Since real roots is not stated in the question, Susan is incorrect.

FACTORING

Factoring: A skill that is used throughout Algebra 2. Depending on the content of a problem, different types of factoring may be required.

Factor Completely:

When an expression or equation is factored completely, each polynomial factor in the answer is prime. ALWAYS look for a Greatest Common Factor (GCF) to factor out; then factor appropriately; finally check all parentheses to make sure further factoring is not possible.

- Factoring is a PROCESS that does *not* mean to solve or find a numerical answer. We can only *solve* when the original expression is an *equation*. Factoring is sometimes used as a method to solve equations or to simplify rational expressions.

- For factoring to result in *x* being equal to something, the expression factored must be part of an *equation* – remember an equation contains =. If the directions simply say "factor" then a value for x, ($x =$ ___) should not be found.

Common Monomial Factor – GCF: (review) A common factor is a

number or a variable, or a product of a number and a variable(s) that is in each term of a polynomial. Determine the greatest numerical factor and the variable(s) with the smallest exponent that is in each term. Write the GCF down, followed by a parenthesis. Inside the parenthesis, write the sum or difference of the quotients of the original terms divided by the GCF. This is often called "factoring the GCF out." The GCF stays as part of the factoring to the end of the problem.

ALWAYS LOOK FOR A GCF BEFORE CONTINUING TO FACTOR.

$2x^2 - 8x + 10$　Number only $2(x^2 - 4x + 5)$

$30x^2y^4 - 39x^2y^2 + 27x^3y^2$　Number and variable(s) with smallest exponent
$3x^2y^2(10y^2 - 13 + 9x)$

$2x^3y^3 - 3x^2y^2z^2 + x^3y^5$　Variables only; no common numerical factor in
the coefficients　$x^2y^2(2xy - 3z^2 + xy^3)$

Algebra 2 Made Easy – Common Core Edition
19

Seeing Structure In Expressions

1.4

Grouping – Common Binomial Factor: Put the polynomial in standard form
If a four term polynomial can be factored by grouping, the product of the
first and fourth terms must be equal to the product of two middle terms.

Examples

❶ $4x^3 + 2x^2 + 2x + 1$ Since $(4x^3)(1) = (2x^2)(2x)$ it can be factored by grouping.

Steps:

1) Put the first two terms into a ( ) as a binomial.
   [*Note:* Add + between the first ( ) and a
   second ( ) that contains the 3rd and 4th terms.] $(4x^3 + 2x^2) + (2x + 1)$

2) Factor a common factor from
   the 1st binomial. $2x^2(2x + 1) + (2x + 1)$

3) Factor a common factor from the
   2nd binomial. If the variable in that term
   is negative, factor out the GCF with a
   negative and factor carefully. $2x^2(2x + 1) + 1(2x + 1)$
   (See example 3.)

4) The expressions that remain in the ( )
   should be alike. The common binomials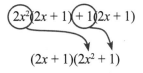
   make one factor and the two original
   GCF's form the 2nd ( ). Combine the $(2x + 1)(2x^2 + 1)$
   two GCF's into the 2nd ( ). Be sure to put
   the + or – sign between them.

5) Multiply to check. $4x^3 + 2x^2 + 2x + 1$ √

❷ $x^3 - 3x^2 + 3x - 9$? *Will it factor?* $(x^3)(-9) = (-3x^2)(3x)$ *Yes*
   $\therefore x^2(x - 3) + 3(x - 3)$
   $(x^2 + 3)(x - 3)$

❸ $2x^3 - 14x^2 - 3x + 21$
   $2x^2(x - 7) - 3(x - 7)$
   $(2x^2 - 3)(x - 7)$

❹ $4x^3 - 12x^2 - x + 3$
   $4x^2(x - 3) - 1(x - 3)$ [*Factor* (–1) *out of the 2nd group.*]
   $(4x^2 - 1)(x - 3)$ [$(4x^2 - 1)$ *is the difference of 2 squares. Factor*]
   $(2x + 1)(2x - 1)(x - 3)$

Binomial Factors: Standard form: $ax^2 + bx + c$

Steps:

1) Make two sets of parenthesis. The factors will be binomials.

2) In the first term of each ( ), a number and the variable will be written. In the 2nd term in each, a number will be written.

3) The product of the coefficients of the first two terms must be a. The product of the last two terms must be c. When these two products are added, they must equal b.

4) When the two binomial factors are multiplied back together, the product is $ax^2 + bx + c$

Note: In some problems the 2nd term in each factor may contain a variable.   For example: $x^2 + 3xy + 2y^2$
$$(x + 2y)(x + y)$$

- When the leading coefficient, $a = 1$: The variable in the factors will have 1 as a coefficient in both parenthesis. The 2nd terms in each ( ) have a product of c and a sum of b as long as both terms are only numbers.

> **Examples**

❶ $x^2 - 3x - 4$ ❷ $x^2 + 8x + 15$ ❸ $x^3 - 3x^2 + 2x$
   $(x - 4)(x + 1)$ $(x + 5)(x + 3)$ $x(x^2 - 3x + 2)$
                                                          $x(x - 2)(x - 1)$

- When a does not $= 1$, use "ARC" factoring. Although binomial factoring can be done by trial and error, or by several other methods, this method works well for my students.

> **Examples**

❶ When the leading coefficient is not 1: $2x^2 + 15x + 18$

Steps:

1) Factor out a GCF if possible. (none here)

2) Multiply $(a)(c)$ $2x^2 + 15x + 18$

3) Rewrite with $a = 1$ and $c = ac$. $x^2 + 15x + 36$ *ARC*

4) Factor the remaining trinomial. $(x + 12)(x + 3)$

5) In both ( ), divide the 2nd terms by the original a. $\left(x + \dfrac{12}{2}\right)\left(x + \dfrac{3}{2}\right)$

6) Simplify if possible. Fractions. $(x + 6)\left(x + \dfrac{3}{2}\right)$

7) If a fraction remains in one or both parentheses, its denominator becomes the coefficient of x in that parenthesis. $(x + 6)(2x + 3)$

9) These are the factors of the original trinomial. $(x + 6)(2x + 3)$

Remember: If a GCF was found, that remains a factor in the final answer.

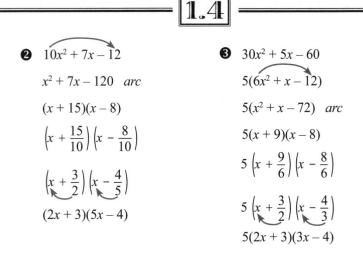

❷ $10x^2 + 7x - 12$

$x^2 + 7x - 120$ *arc*

$(x + 15)(x - 8)$

$\left(x + \dfrac{15}{10}\right)\left(x - \dfrac{8}{10}\right)$

$\left(x + \dfrac{3}{2}\right)\left(x - \dfrac{4}{5}\right)$

$(2x + 3)(5x - 4)$

❸ $30x^2 + 5x - 60$

$5(6x^2 + x - 12)$

$5(x^2 + x - 72)$ *arc*

$5(x + 9)(x - 8)$

$5\left(x + \dfrac{9}{6}\right)\left(x - \dfrac{8}{6}\right)$

$5\left(x + \dfrac{3}{2}\right)\left(x - \dfrac{4}{3}\right)$

$5(2x + 3)(3x - 4)$

Factoring the Difference of Two Perfect Squares: A binomial in the form of $(ax)^2 - c^2$ is factored as $(ax + c)(ax - c)$

Note: $(ax + c)$ and $(ax - c)$ are called conjugates of each other. (See page 10.)

 Examples

❶ $x^2 - 49$
$(x + 7)(x - 7)$

❷ $25x^2 - 36$
$(5x + 6)(5x - 6)$

❸ $27x^2 - 75$
$3(9x^2 - 25)$
$3(3x + 5)(3x - 5)$

Check in case further factoring is possible.

❹ $16x^4 - y^4$
$(4x^2 - y^2)(4x^2 + y^2)$
$(2x + y)(2x - y)(4x^2 + y^2)$

Factoring the Sum or Difference of Two Perfect Cubes: There are formulas to use for this type of factoring.

$x^3 + n^3 = (x + n)(x^2 - nx + n^2)$: $x^3 + 27 = (x + 3)(x^2 - 3x + 9)$

$x^3 - n^3 = (x - n)(x^2 + nx + n^2)$: $x^3 - 27 = (x - 3)(x^2 + 3x + 9)$

If x^3 has a coefficient, its cube root must be included with x:

Example $8x^3 - 27 = (2x)^3 - (3)^3 = ((2x) - 3)((2x)^2 + 3(2x) + 3^2)$
$(2x - 3)(4x^2 + 6x + 9)$

Examples

❶ Determine the value of n if $(x-2)^n(x+2)$ is equivalent to $x^2(x-2) - 4(x-2)$.

Solution: Factor the given expression and rewrite in simplest form to find the value of n.
$x^2(x-2) - 4(x-2) \Rightarrow (x^2-4)(x-2) \Rightarrow (x-2)(x+2)(x-2)$
or $(x-2)^2(x+2)$
Therefore, $n = 2$

❷ Write all the factors of $x^4 - 4x^2 - 45$ when factored completely.

Solution: $x^4 - 4x^2 - 45 \Rightarrow (x^2-9)(x^2+5)$
$(x+3)(x-3)(x^2+5)$

❸ Determine all the factors of the expression $2x^3 + 6x^2 - 8x - 24$ if one of the factors is $(x+3)$.

Solution: This can be factored by grouping since the product of the first and last terms is equal to the product of the middle two terms.
$(2x^3)(-24) = (6x^2)(-8x) = 48x^2$
$2x^3 + 6x^2 - 8x - 24 \ 2(x^3 + 3x^2 - 4x - 12)$
$2[x^2(x+3) - 4(x+3)] \Rightarrow 2[(x^2-4)(x+3)]$
$2[(x+2)(x-2)(x+3)]$

❹ Use the graph of $f(x) = x^3 + x^2 - 6x$ to write the linear factors of the function.

Solution: Make a graph and determine the zeroes. The factors will contain the numbers opposite the zeroes of the function. The zeroes are at $x = -3$, $x = 0$, and $x = 2$. The factors are $(x+3)(x)(x-2)$ which would usually be written $x(x+3)(x-2)$.

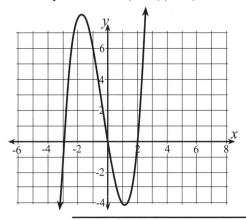

Seeing Structure In Expressions

POLYNOMIALS

A polynomial can be written in the form $a_n x^n = a^{n-1} x^{n-1} + \ldots + a_2 x^2 + a_0$, where n is a nonnegative integer and a_i is a real number. The factors of a polynomial function can be used to sketch a graph of the function since the zeros of the function are closely related to the zeros of the function. If a factor is set equal to zero and solved, the value of x obtained is a zero of the function. When one polynomial is divided by another, it is a rational expression. A polynomial division problem is a rational expression.

REMAINDER THEOREM

Remainder Theorem: For a polynomial, $p(x)$, and a number, a, the remainder on division by $x - a$ is $p(a)$, so $p(a) = 0$ if and only if $(x - a)$ is a factor of $p(x)$.

In simpler language, if a polynomial is divisible by a true factor, $(x - a)$, of that polynomial, the remainder is zero. The zero remainder also indicates that a is a root or zero of the given polynomial. The Remainder Theorem can be used to determine if a given factor is actually a factor of a polynomial or if a given number is a root or zero of a polynomial. Two methods are commonly used with the Remainder Theorem, long division and synthetic division.

Long Division: Long division of a polynomial by another polynomial is very similar to long division of large numbers. (See example 1 on page 25.)

Synthetic Division: A shorter method to divide polynomials. (See examples 1 and 2 on pages 26-27.)

Note: See additional Remainder Theorem examples on page 27-29.

LONG DIVISION

Examples

Arithmetic with Polynomials and Rational Expressions

❶ Determine if $(x - 4)$ is a factor of $p(x) = x^3 - 5x - 6$.
In this example, $a = 4$.

Steps:

1) Set up the division example with $(x - 4)$ as the divisor.

2) Make room for the missing x^2 term when writing the dividend. Use $0x^2$ as a placeholder.

3) Divide the 1st term of the dividend by the 1st term of the divisor and put the quotient above the 1st term in the dividend. $x^3 \div x = x^2$

4) Multiply the divisor by the quotient and write the product under the dividend lining up the like terms. $(x - 4)(x^2) = x^3 - 4x^2$

5) Subtract. $x^3 + 0x^2 - (x^3 - 4x^2) = 4x^2$

6) Bring down the next term in the dividend.

7) Repeat the process from steps 3-6.

8) If the final difference is not 0, the difference is the remainder. The remainder here is 38.

$$x^2 + 4x + 11 + \frac{38}{x - 4}$$

$$x - 4 \overline{)\; x^3 + 0x^2 - 5x - 6}$$
$$-(x^3 - 4x^2)$$
$$4x^2 - 5x$$
$$-(4x^2 - 16x)$$
$$11x - 6$$
$$-(11x - 44)$$
$$+38$$

9) The remainder can be written as a fraction with the divisor as the denominator and added to the quotient.

Conclusion: The remainder here is not zero. Therefore, $(x - 4)$ is not a factor of $p(x)$ and 4 is not a zero of $p(x)$.

❷ Is $(x - 2)$ a factor of $h(x) = 2x^3 - 3x^2 - 5x + 6$?

$$x - 2 \overline{)\; 2x^3 - 3x^2 - 5x + 6}$$
$$\begin{array}{r} 2x^2 + x - 3 \\ \hline \end{array}$$
$$-(2x^3 - 4x^2)$$
$$x^2 - 5x$$
$$-(x^2 - 2x)$$
$$-3x + 6$$
$$-(-3x + 6)$$
$$0$$

Conclusion: Since the remainder is zero after dividing $h(x)$ by $(x - 2)$, $(x - 2)$ is an actual factor of $h(x)$. Because $(x - 2)$ is a factor of $h(x)$, $x = 2$ is a zero of $h(x)$.

SYNTHETIC DIVISION

A shortcut method to divide a polynomial by a binomial is called synthetic division. Be careful because it requires a very different procedure than long division does. A "skeleton" of the problem is set up with only numbers, no variables. The division is done, and then the variables can be inserted back into the quotient.

Examples Synthetic Division

❶ Using the same function and linear factor as we used in example 1 on the previous page, this is what synthetic division will look like. $p(x) = x^3 - 5x - 6$ divided by $(x - 4)$.

Steps:

1) Make an upside down division sign – or a box with 2 sides as shown below.

2) Inside the box, put the coefficients of the terms of the polynomial – if a term is missing make its coefficient a 0. Outside the box, put the value of a which is the number that is being tested to see if it is a zero of the function. $a = 4$ since $(x - 4)$ is the factor given.

$$4 \,\begin{array}{|cccc} 1 & 0 & -5 & -6 \\ \hline & & & \end{array}$$
$$\;1$$

3) Carry down the first coefficient inside the box, put it under the line.

- -

4) Multiply the number carried down by the number outside the box. (the test number, a). Put the product inside the box, in the next column.

$$4 \,\begin{array}{|cccc} 1 & 0 & -5 & -6 \\ & 4 & & \\ \hline 1 & & & \end{array}$$

- -

5) ADD row 1 and row 2. Put the sum under the box in that column. $0 + 4 = 4$

$$4 \,\begin{array}{|cccc} 1 & 0 & -5 & -6 \\ & 4 & & \\ \hline 1 & 4 & & \end{array}$$

- -

6) Multiply the sum from step 6 by a and place it in the next column inside the box. $4 \cdot 4 = 16$

7) Continue steps 4-7 as needed.

$$4 \,\begin{array}{|cccc} 1 & 0 & -5 & -6 \\ & 4 & 16 & 44 \\ \hline 1 & 4 & 11 & 36 \end{array}$$

- -

8) The sum under the last column will be a number or zero. If it is a number, that is the remainder of the division. If it is zero, then the factor that was tested $(x - a)$ is an actual factor of the function in which case a zero of the function would be the value of a.

Conclusion: The sum in the final column of this example is 36. Therefore $(x - 4)$ is not a factor of $p(x) = x^3 - 5x - 6$ and 4 is not a zero of $p(x)$.

Algebra 2 Made Easy – Common Core Edition

❷ Determine if $(x - 2)$ is a factor of $h(x) = 2x^2 - 3x^2 - 5x + 6$.
Is 2 a zero of $p(x)$?

Solution: Use synthetic division to test and see if 2 is a zero of
the $p(x)$. If it is, then $(x - 2)$ is a factor of $h(x)$.

$$
\begin{array}{c|cccc}
2 & 2 & -3 & -5 & 6 \\
  &   & 4 & 2 & -6 \\
\hline
  & 2 & 1 & -3 & 0
\end{array}
$$

Conclusion: Testing the number 2 resulted in a remainder of zero in
synthetic division. That indicates that 2 is a zero of $h(x)$ and $(x - 2)$
is a factor of $h(x)$.

Extra Examples – Remainder Theorem

❶ What is the rational expression $\dfrac{x^2 + 67x - 40}{x - 5}$ equivalent to?

Solution: Divide. $\dfrac{x^2 + 7x - 40}{x - 5} = (x - 5) \overline{)x^2 + 7x - 40}$

$$
\begin{array}{r}
x + 12 + \dfrac{20}{x-5} \\
x - 5 \overline{)x^2 + 7x - 40} \\
-(x^2 - 5x) \\
\hline
12x - 40 \\
-(12x - 60) \\
\hline
20
\end{array}
$$

❷ Use the Remainder Theorem to determine whether $(x + 3)$ is a factor
of $f(x) = x^5 + 3x^4 - 3x^3 - 6x^2 - 3x - 12$. Explain what this information
reveals about the zeros of $f(x)$ and to the graph of $f(x)$.

Solution: Long division can be used here but this is a very long
problem. Synthetic division will be more efficient.

$$
\begin{array}{c|cccccc}
-3 & 1 & +3 & -3 & -6 & -3 & -12 \\
   &   & -3 & 0 & 9 & -9 & 36 \\
\hline
   & 1 & 0 & -3 & 3 & -12 & 24
\end{array}
$$

Since the number in the final step of the problem is 24, $(x + 3)$ is not a
factor of $f(x)$. Therefore, -3 is not a zero of $f(x)$ and the graph of $f(x)$ does
not intersect the x axis at -3.

❸ What is the remainder if the function graphed here is divided by $(x + 1)$?

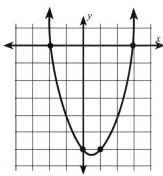

Solution: If $(x + 1)$ were a factor of the function, then -1 would be a zero of the function and the value of the function at -1 would be 0. The value of this function at $x = -1$ is -4.

Conclusion: The remainder when the function is divided by $(x + 1)$ is -4.

Alternate Method: The roots or zeros of this function can be read from the graph as $x = -2$ and $x = 3$, making the factors of the function $(x + 2)(x - 3)$. Multiply the factors to find the function, then use the Remainder Theorem to determine if $(x + 1)$ is a factor. Use long division or synthetic division to find the remainder when the function is divided by $(x + 1)$.

$$(x + 2)(x - 3) = x^2 - x - 6$$

$$
\begin{array}{r|rrr}
-1 & 1 & -1 & -6 \\
   &   & -1 & 2 \\
\hline
   & 1 & -2 & \boxed{-4}
\end{array}
$$

Conclusion: The remainder is -4.

❹ On her final exam Julie is given the following problem and she has no clue about how to solve it. Describe a method she could use to determine the correct answer and explain your reasoning.

Question: If $h(x) = 2x^3 - bx^2 - 22x - 15$, find the value of b if $(x + 5)$ is a factor of $h(x)$. Rewrite $h(x)$ with the appropriate value of b shown.

Solution: Since $(x + 5)$ is a factor of $h(x)$, a zero of the function is -5. By substituting -5 in the function for x, and making the function $= 0$, the value of b can be found.

$$h(x) = 2x^3 - bx^2 - 22x - 15$$
$$0 = 2(-5)^3 - b(-5)^2 + 22(-5) - 15$$
$$0 = -25b - 375$$
$$25b = -375$$
$$b = -15$$

Conclusion: The value of b is -15. The function rewritten with b is $h(x) = 2x^3 - (-15)x^2 - 22x - 15$ *or* $h(x) = 2x^3 + 15x^2 - 22x - 15$

Algebra 2 Made Easy – Common Core Edition

❺ Find the zeros of $p(x) = (x^2 - 4)(x^2 + 4)$.

Solution: At the zeros, the value of $p(x) = 0$. Use the Zero Product Property and set each factor equal to zero and solve.

$(x^2 - 4)(x^2 + 4) = 0$

$(x^2 - 4) = 0$	$(x^2 + 4) = 0$
$x = \pm\sqrt{4}$	$x = \pm\sqrt{-4}$
$x = 2, \ x = -2$	$x = 2i, \ x = -2i$

Conclusion: The real roots or zeros of $p(x)$ are 2 and –2. There are also two imaginary zeros, $2i$ and $-2i$.

❻ **Using long division:**

  a) find $q(x)$, the quotient of $\dfrac{a(x)}{b(x)}$.

  b) Express the remainder of $\dfrac{a(x)}{b(x)}$ in the form $\dfrac{r(x)}{b(x)}$

  where $r(x) < b(x)$ and $r(x)$ is the remainder.

Given: $a(x) = -6x^3 + 10x^2 + 15x - 4$ and $b(x) = 2x^2 + x + 3$

Solution: $\quad -3x + 6 + \dfrac{x^2 - 22}{2x^2 + x + 3}$

$$
\begin{array}{r}
2x^2 + x + 3 \overline{\smash{)}\,-6x^3 + 10x^2 + 15x - 4} \\
-(-6x^3 - 3x^2 - 9x) \\
\hline
13x^2 + 6x - 4 \\
-(12x^2 + 6x + 18) \\
\hline
x^2 \qquad -22
\end{array}
$$

Conclusion:

a) The quotient, $q(x)$, is $-3x + 6$ with a remainder of $x^2 - 22$.

b) The remainder is written $\dfrac{x^2 - 22}{2x^2 + x + 3}$.

Note: $\dfrac{-6x^3 + 10x^2 + 15x - 4}{2x^2 + x + 3}$ is a rational expression.

Arithmetic with Polynomials
and Rational Expressions

POLYNOMIAL IDENTITIES

An identity is an equation that is true and it can be generalized to be used in a variety of situations. In order to prove an identity is true, choose one side of the equation and use accepted algebraic procedures to match that side to the other side of the equation. It is sometimes necessary to work on both sides of the equation, but commonly one side is developed to match the other. Expansion of a term and use of the distributive property are often used in the proof.

A simple numerical example of an identity is $6^2 = 36$. The proof could be:

$$6^2 = 36 \qquad\qquad\qquad 6^2 = 36$$
$$(6)(6) = 36 \qquad\qquad 6^2 = (9)(4)$$
$$36 = 36 \qquad or \qquad 6^2 = (3^2)(2^2)$$
$$6^2 = (3 \cdot 2)^2$$
$$6^2 = 6^2$$

Polynomial Identities and their proofs include:

1) THE SQUARE OF A BINOMIAL

$$(a + b)^2 = a^2 + 2ab + b^2$$
$$(a + b)(a + b) = a^2 + 2ab + b^2$$
$$a^2 + ab + b^2 = a^2 + 2ab + b^2$$
$$a^2 + 2ab + b^2 = a^2 + 2ab + b^2$$

Example:

$$(x + 5)^2 = x^2 + 2(1)(5)x + 25$$
$$(x + 5)^2 = x^2 + 10x + 25$$

2) THE PRODUCT OF A BINOMIAL

$$(a + b)(c + d) = ac + ad + bc + bd$$
$$ac + ad + bc + bd = ac + ad + bc + bd$$

Example:

$$(x + 4)(y - 3) = xy - 3x + 4y - 12$$

3) THE DIFFERENCE OF TWO PERFECT SQUARES

$$a^2 - b^2 = (a + b)(a - b)$$
$$a^2 - b^2 = a^2 - ab + ab - b^2$$
$$a^2 - b^2 = a^2 - b^2$$

Example:

$$16x^2 - 81y^2 = (4x + 9y)(4x - 9y)$$
$$a = 4x \qquad b = 9y$$

4) THE FACTORS OF A TRINOMIAL

$$x^2 + (a + b)x + ab = (x + a)(x + b)$$
$$x^2 + (a + b)x + ab = x^2 + bx + ax + ab$$
$$x^2 + (a + b)x + ab = x^2 + (a + b)x + ab$$

Example:

$$x^2 + 7x + 12 = (x + 3)(x + 4)$$

Algebra 2 Made Easy – Common Core Edition

5) THE SUM OF TWO CUBES

   Example:

   $a^3 + b^3 = (a + b)(a^2 - ab + b^2)$

   $8x^2 + 27 = (2x + 3)(4x^2 - 6x + 9)$

   $a^3 + b^3 = a^3 - a^2b + ab^2 + a^2b - ab^2 + b^3$

   $a = 2x \qquad b = 3$

   $a^3 + b^3 = a^3 + b^3$

6) THE DIFFERENCE OF TWO CUBES

   Example:

   $a^3 - b^3 = (a - b)(a^2 + ab + b^2)$

   $125x^3 - 8 = (5x - 2)(25x^2 + 10x + 4)$

   $a^3 - b^3 = a^3 + a^2b + ab^2 - a^2b - ab^2 - b^3$

   $a = 5x \qquad b = 2$

   $a^3 - b^3 = a^3 - b^3$

7) PYTHAGOREAN IDENTITY: THE PYTHAGOREAN IDENTITY
   CAN BE USED TO GENERATE PYTHAGOREAN TRIPLES.
   THE IDENTITY IS $(x^2 - y^2) + (2xy)^2 = (x^2 + y^2)^2$

$$(x^2 + y^2)^2 = (x^2 - y^2) + (2xy)^2$$

$$(x^2 + y^2)(x^2 + y^2) = (x^2 - y^2) + (2xy)^2$$

$$x^4 + x^2y^2 + x^2y^2 + y^4 = (x^2 - y^2) + (2xy)^2$$

$$x^4 + 2x^2y^2 + y^4 = (x^2 - y^2) + (2xy)^2$$

$$x^4 - 2x^2y^2 + x^2y^2 + y^4 = (x^2 - y^2) + (2xy)^2$$

$$(x^4 - 2x^2y^2 + y^4) + 4x^2y^2 = (x^2 - y^2) + (2xy)^2$$

Example: Generate a Pythagorean Triple if $x = 7$ and $y = 12$.

$$a^2 - b^2 = c^2$$

$$(7^2 - 12^2)^2 + [2(7)(12)] = (7^2 + 12^2)^2$$

$a^2 = (7^2 - 12^2)$	$b^2 = (2xy)^2$	$c^2 = (7^2 + 12^2)^2$
$a^2 = (49 - 144)^2$	$b^2 = [2(7)(12)]^2$	$c^2 = (49 + 144)^2$
$a = \sqrt{(9025)}$	$b = \sqrt{28224}$	$c = \sqrt{37249}$
$a = 95$	$b = 168$	$c = 193$

Conclusion: When $x = 7$ and $y = 12$, the Pythagorean Triple generated is 95, 168, and 193.

8) THE QUADRATIC FORMULA PROOF USES THE METHOD OF COMPLETING THE SQUARE. (See page 37.)

$$if \quad : \quad ax^2 + bx + c = 0, \text{ then } x = \frac{-b \pm \sqrt{b^2 - 4ac}}{2a}$$

$$ax^2 + bx + c = 0$$

Complete the square
$$\begin{cases} x^2 + \dfrac{bx}{a} + \dfrac{c}{a} = 0 \\[2mm] x^2 + \dfrac{bx}{a} = -\dfrac{c}{a} \\[2mm] x^2 + \dfrac{bx}{a} + \left(\dfrac{b}{2a}\right)^2 = \left(\dfrac{b}{2a}\right)^2 - \dfrac{c}{a} \end{cases}$$

Solve for x
$$\begin{cases} \left(x + \dfrac{b}{2a}\right)^2 = \left(\dfrac{b}{2a}\right)^2 - \dfrac{c}{a} \\[2mm] x + \dfrac{b}{2a} = \pm\sqrt{\left(\dfrac{b}{2a}\right)^2 - \dfrac{c}{a}} \\[2mm] x = -\dfrac{b}{2a} \pm \sqrt{\left(\dfrac{b}{2a}\right)^2 - \dfrac{c}{a}} \\[2mm] x = -\dfrac{b}{2a} \pm \sqrt{\dfrac{b^2}{4a^2} - \dfrac{c}{a}} \\[2mm] x = \dfrac{-b}{2a} \pm \sqrt{\dfrac{b^2 - 4ac}{4a^2}} \end{cases}$$

Simplify : $x = \dfrac{-b}{2a} \pm \dfrac{\sqrt{b^2 - 4ac}}{2a}$ or $x = \dfrac{-b \pm \sqrt{b^2 - 4ac}}{2a}$

Example: $x^2 - 10x + 7 = 0$

$$a = 1, \; b = -10, \; c = 7$$

$$x = \frac{-b \pm \sqrt{b^2 - 4ac}}{2a}$$

$$x = \frac{-(-10) \pm \sqrt{(-10)^2 - 4(1)(7)}}{2(1)}$$

$$x = \frac{10 \pm \sqrt{100 - 28}}{2}$$

$$x = \frac{10 \pm \sqrt{72}}{2}$$

$$x = \frac{10 \pm 6\sqrt{2}}{2}$$

$$x = 5 \pm 3\sqrt{2}$$

Algebra 2 Made Easy – Common Core Edition

Examples

❶ Juan uses the Polynomial Identity $(x^2 - y^2)^2 + (2xy)^2 = (x^2 + y^2)^2$ to generate the Pythagorean Triple 21, 20, 29. What values of x and y did he use?

Solution: Use the Pythagorean Theorem, $c^2 + b^2 = c^2$ in place of the identity to find x and y.

$a^2 = (x^2 - y^2)^2$ $b^2 = (2xy)^2$ $c^2 = (x^2 + y^2)^2$
$a^2 = 21$ $b^2 = 20$ $c^2 = 29$
$(21)^2 = (x^2 - y^2)^2$ $20^2 = (2xy)^2$ $c^2 = 2^2$
$21 = x^2 - y^2$ $20 = 2xy$ $c^2 = x^2 + y^2$

Choose 2 of the equations above and solve them simultaneously by adding. Substitute in the third equation to find y.

$21 = x^2 - y^2$ $20 = 2xy$
$29 = x^2 + y^2$ $20 = 2(5)y$
$50 = 2x^2$ $y = 4$
$x^2 = 25$
$x = 5$

Conclusion: Juan used $x = 5$ and $y = 4$ to generate the Pythagorean Triples 21, 20, 29.

❷ Use the following identity to simplify $(3n - 8)^3$. Explain your reasoning.
$$(x + y)^3 = x^3 + 3x^2y + 3xy^2 + y^3$$

Solution: In this example, x is $3n$, and y is -8. Substitute those values for x and y in the identity and simplify.
$(3n - 8)^3 = (3n)^3 + 3(3n)^2(-8) + 3(3n)(-8)^3 + (-8)^3$
$(3n - 8)^3 = 27n^3 + 648n^2 - 4608n - 512$

❸ The point $\left(\dfrac{12}{13}, \dfrac{5}{13}\right)$ is point on the circle $x^2 + y^2 = 1$. Use the equation of a circle to prove this is a point on the circle.

Solution: Substitute $\dfrac{12}{13}$ for x and $\dfrac{5}{13}$ for y in the circle equation. Simply. If the left side of the equation is equal to one, it is a point on the circle.

$\left(\dfrac{12}{13}\right)^2 + \left(\dfrac{5}{13}\right)^2 = 1$

$\dfrac{144}{169} + \dfrac{25}{169} = 1$

$\dfrac{169}{169} = 1$ **Conclusion:** The point $\left(\dfrac{12}{13}, \dfrac{5}{13}\right)$ is on the circle $x^2 + y^2 = 1$

$1 = 1$

Arithmetic with Polynomials and Rational Expressions

1.6

QUADRATIC EQUATIONS

A quadratic equation written in standard form is $ax^2 + bx + c = 0$ where $a \neq 0$. Quadratic equations can be solved in several ways and most methods require that the equation be written in standard form. If the equation is not given in this form, change it to standard form before beginning to work on it. Examine the equation to determine a solving method. Quadratic equations can be solved algebraically or graphically.

Note: In certain types of problems, one or more of the solutions must be rejected due to conditions in the problem.

Examples

①
$$x^2 - 2x = 4$$
Rewrite: $\qquad x^2 - 2x - 4 = 0$
In this example: $\qquad a = 1, b = -2,$ and $c = -4$

②
$$3x^2 = 15$$
Rewrite: $\qquad 3x^2 - 15 = 0$
In this example: $\qquad a = 3, \ b = 0, \ $ and $c = -15$

SOLVING QUADRATIC EQUATIONS

- **Factor to Solve:** If a quadratic equation can be factored the zero product property can be used to find the solutions. For all types of factoring equations, factor out the greatest common factor first if there is one. (See Unit 1.4 – Factoring.)

    Steps:

    1) Factor.

    2) Make each factor = 0. (Example 1)

        a) If the GCF is a number, divide both sides of the equation by that number and remove it from the problem. (Example 2)

        b) If the GCF was a variable, it is also a factor so make that variable = 0, as well as making the other factor = 0. (Example 3)

    3) Solve both equations.

    4) Each solution must be checked. – Substitute each value of *x* into the original equation and see if each solution checks.

Examples

❶ $x^2 - 2x = 3$

$x^2 - 2x - 3 = 0$

$(x - 3)(x + 1) = 0$

$x - 3 = 0 \quad x + 1 = 0$

$x = 3 \quad x = -1$

CHECK

$x^2 - 2x = 3$

$(3)^2 - 2(3) \overset{?}{=} 3; \quad 9 - 6 = 3; \quad 3 = 3 \checkmark$

$(-1)^2 - 2(-1) \overset{?}{=} 3; \quad 1 + 2 = 3; \quad 3 = 3 \checkmark$

Solution = $\{-1, 3\}$

❷ $2x^2 + 10x - 12 = 0$

$2(x^2 + 5x - 6) = 0$

$\dfrac{2\left(x^2 + 5x - 6\right)}{2} = \dfrac{0}{2}$

$(x^2 + 5x - 6) = 0$

$(x + 6)(x - 1) = 0$

$x + 6 = 0 \quad x - 1 = 0$

$x = -6 \quad x = 1$

CHECK

$2x^2 + 10x - 12 = 0$

$2(-6)^2 + 10(-6) - 12 = 0$

$72 - 60 - 12 = 0; \quad 0 = 0 \checkmark$

$2(1)^2 + 10(1) \overset{?}{=} 12 = 0$

$2 + 10 - 12 \overset{?}{=} 0; \quad 0 = 0 \checkmark$

Solution = $\{-6, 1\}$

❸ $x^2 - 5x = 0$

$x(x - 5) = 0$

$x = 0 \quad x - 5 = 0$

$x = 0 \quad x = 5$

CHECK

$x^2 - 5x = 0$

$0^2 - 5(0) \overset{?}{=} 0; \quad 0 = 0 \checkmark$

$5^2 - 5(5) \overset{?}{=} 0; \quad 0 = 0 \checkmark$

Solution $\overset{?}{=} \{0, 5\}$

- Use **square roots** if $b = 0$ to find the positive and negative square roots of x^2.
   Steps:
   1) Rearrange so $x^2 = \dfrac{-c}{a}$

   2) Solve for x: $\pm\sqrt{\dfrac{-c}{a}}$. Leave the answer in simplified radical form or round as directed.

   3) Check both answers in the original equation. Checks are not shown in the examples.

Examples

❶ $2x^2 - 49 = 0$

$x^2 = \dfrac{49}{2}$

$x = \pm\sqrt{\dfrac{49}{2}}$

$x = \pm\dfrac{7}{\sqrt{2}} \bullet \dfrac{\sqrt{2}}{\sqrt{2}} = \pm\dfrac{7\sqrt{2}}{2}$

❷ $x^2 + 50 = 0$

$x^2 = -50$

$x = \pm\sqrt{-50}$

$x = \pm 5i\sqrt{2}$

❸ $x^2 - 16 = 0$

$x^2 = 16$

$x = \pm\sqrt{16} = \pm 4$

❹ $x^2 - 20 = 0$

$x^2 = 20$

$x = \pm\sqrt{20} = \pm 2\sqrt{5}$

Arithmetic with Polynomials and Rational Expressions

- Solve using the **Quadratic Formula:** $x = \dfrac{-b \pm \sqrt{b^2 - 4ac}}{2a}$

  Steps:
  1) Identify a, b, and c.
  2) Substitute the numbers into the formula.
  3) "Do the math." Perform the indicated operations and simplify. Be careful with the signs.
  4) Leave in simplest radical form, $a + bi$ form, or as a decimal using the calculator if required.

Examples

❶ $3x^2 - 5x + 2 = 0$

$a = 3$, $b = (-5)$, $c = 2$

$$x = \frac{-(-5) \pm \sqrt{(-5)^2 - 4(3)(2)}}{2(3)}$$

$$x = \frac{5 \pm \sqrt{25 - 24}}{6}$$

$$x = \frac{5 + 1}{6} = 1, \quad x = \frac{5 - 1}{6} = \frac{2}{3}$$

Solution : $x = \{1, \frac{2}{3}\}$

❷ $4x^2 - 4x - 11 = 0$

$a = 4$, $b = (-4)$, $c = (-11)$

$$x = \frac{-(-4) \pm \sqrt{(-4)^2 - 4(4)(-11)}}{2(4)}$$

$$x = \frac{4 \pm \sqrt{16 + 176}}{8} = \frac{4 \pm \sqrt{192}}{8}$$

$$x = \frac{4 + 8\sqrt{3}}{8}, \quad x = \frac{4 - 8\sqrt{3}}{8}$$

$$x = \frac{1 + 2\sqrt{3}}{2}, \quad x = \frac{1 - 2\sqrt{3}}{2}$$

Solution : $x = \left\{ \dfrac{1 + 2\sqrt{3}}{2}, \ \dfrac{1 - 2\sqrt{3}}{2} \right\}$

❸ $x^2 - 2x + 6 = 0$

$a = 1$, $b = (-2)$, $c = 6$

$$x = \frac{-(-2) \pm \sqrt{(-2)^2 - 4(1)(6)}}{2(1)}$$

$$x = \frac{2 \pm \sqrt{4 - 24}}{2}$$

$$x = \frac{2 \pm \sqrt{-20}}{2}$$

$$x = \frac{2 \pm i\sqrt{20}}{2} = \frac{2}{2} \pm \frac{2i\sqrt{5}}{2}$$

$$x = 1 + i\sqrt{5}, \quad x = 1 - i\sqrt{5}$$

$$SS = \left\{ 1 + i\sqrt{5}, \ 1 - i\sqrt{5} \right\}$$

Algebra 2 Made Easy – Common Core Edition

- If $a = 1$, solve by **completing the square**. If $a \neq 1$, divide each term in the equation by a before beginning. It is often more efficient to use the quadratic formula when $a \neq 1$.

Examples

❶ $x^2 - 6x + 2 = 0$

Steps

1) Subtract c from both sides. $\qquad x^2 - 6x = -2$

2) Divide : b by 2 and square it. $\qquad b = -6 : \left(\dfrac{-6}{2}\right)^2 = (-3)^2 = 9$

3) Add to BOTH sides of the equation. $\qquad x^2 - 6x + 9 = -2 + 9$

4) Factor the left side into 2 equal factors. $\quad (x - 3)(x - 3) = 7$

5) Write as binomial squared. $\qquad (x - 3)^2 = 7$

6) Take the square roots of both sides. $\qquad (x - 3) = \pm\sqrt{7}$

7) Solve for x. $\qquad x = 3 + \sqrt{7}, \quad x = 3 - \sqrt{7}$

Solution: $x = \{3 + \sqrt{7}, 3 - \sqrt{7}\}$

❷ $x^2 - 3x + 1 = 0$

$$x^2 - 3x = -1$$

$$b = (-3) : \left(\dfrac{-3}{2}\right)^2 = \left(\dfrac{9}{4}\right)$$

$$x^2 - 3x + \dfrac{9}{4} = -1 + \dfrac{9}{4}$$

$$\left(x - \dfrac{3}{2}\right)^2 = \dfrac{5}{4}$$

$$x - \dfrac{3}{2} = \pm\sqrt{\dfrac{5}{4}}$$

$$x = \dfrac{3}{2} + \dfrac{\sqrt{5}}{2} = \dfrac{3 + \sqrt{5}}{2}; \quad x = \dfrac{3}{2} - \dfrac{\sqrt{5}}{2} = \dfrac{3 - \sqrt{5}}{2}$$

Solution : $x = \dfrac{3 + \sqrt{5}}{2}$ and $x = \dfrac{3 - \sqrt{5}}{2}$

Arithmetic with Polynomials and Rational Expressions

1.6

THE DISCRIMINANT

Discriminant: A part of the quadratic formula which allows us to anticipate what kinds of solutions (roots) a quadratic equation will have.

The formula is $b^2 - 4ac$:

Steps:
  1) Write the equation in standard form.
  2) Identify a, b, and c.
  3) Substitute them in the discriminant: $b^2 - 4ac$.
  4) Evaluate.
  5) Determine the nature of the roots (type of roots) based on these rules:

Description of the value of the discriminant: $b^2 - 4ac$	Nature or type of the roots	Number of roots – number of x-axis intercepts
> 0, a perfect square	real, unequal, rational	2
> 0, not a perfect square	real, unequal, irrational	2
< 0	No real roots	0
= 0	Real, rational, equal	1 (also called a double root)

Why does this help us? By evaluating the discriminant, we know whether to factor or use the quadratic equation. We can answer questions about the equation and the graph of the equation. We can determine before solving whether the root(s) are real or complex.

Examples   Use the discriminant, $b^2 - 4ac$, to answer the following questions about roots.

❶   If the roots are $x = -3$ and $x = 7$, what could the discriminant be?
     Answer: Greater than zero, a perfect square.

❷   If the discriminant = 15, describe the roots.
     Answer: There are two roots that are real, unequal, and irrational.

❸   Find a value of b so the roots of $x^2 + bx + 5 = 0$ are
   a) **Equal:** The discriminant must = 0.

$$b^2 - 4(1)(5) = 0$$
$$b^2 = 20$$
$$b = \pm\sqrt{20} = \pm 2\sqrt{5}$$
$$b = 2\sqrt{5} \text{ or } b = -2\sqrt{5}$$

     The equation is $x^2 + 2\sqrt{5}x + 5 = 0$ *or* $x^2 - 2\sqrt{5}x + 5 = 0$

   b) **Not real numbers:** The discriminant must be less than zero.

$$b^2 - 4ac < 0$$
$$b^2 - 4(1)(5) < 0$$
$$b^2 < 20; \quad b < 2\sqrt{5} \text{ and } b > -2\sqrt{5}$$
$$-2\sqrt{5} < b < 2\sqrt{5}$$

     Any value between $-2\sqrt{5}$ and $2\sqrt{5}$ will have complex roots.

Algebra 2 Made Easy – Common Core Edition

④ Identify the graph of a quadratic equation based on the value of its discriminant.

To solve, look at the graph.

a) If the graph intersects the *x*-axis in two different places, the discriminant is a positive real number.

b) If the graph is tangent to the *x*-axis at one point, the discriminant is 0.

c) If the graph does not intersect the *x*-axis, the discriminant is negative.

Describe the nature of the roots. Then solve each equation.

Examples

❶ $x^2 - 4x + 2 = 0$

$a = 1, \ b = (-4), \ c = 2$

$b^2 - 4ac = (-4)^2 - 4(1)(2) = 8$

Since 8 is greater than zero and is not a perfect square, the roots should be two unequal, irrational, real numbers.

Since we know it has irrational roots, factoring won't work. Use the quadratic formula (or completing the square).

$x^2 - 4x + 2 = 0$

$a = 1, \ b = (-4), \ c = 2$

$x = \dfrac{-b \pm \sqrt{b^2 - 4ac}}{2a}$

$x = \dfrac{-(-4) \pm \sqrt{(-4)^2 - 4(1)(2)}}{2(1)}$

$x = \dfrac{4 \pm \sqrt{8}}{2}$

$x = \dfrac{4 \pm 2\sqrt{2}}{2}$

$x = 2 + \sqrt{2}, \quad x = 2 - \sqrt{2}$

Two real, unequal and irrational roots.

❷ $x^2 - 3x - 4 = 0$

$a = 1$, $b = (-3)$, $c = (-4)$

$b^2 - 4ac = (-3)^2 - 4(1)(-4) = 25$

25 is a perfect square and is greater than zero. There should be two real, unequal, rational roots. It can be factored.

$(x - 4)(x + 1) = 0$

$x - 4 = 0 \qquad x + 1 = 0$

$\qquad x = 4 \qquad\qquad x = -1$

Two real, unequal, rational roots

❸ $x^2 - 4x + 7 = 0$

$a = 1$, $b = (-4)$, $c = 7$

$b^2 - 4ac = (-4)^2 - 4(1)(7) = -12$

The discriminant is less than zero, so there should be no real roots. Use the quadratic formula.

$$x = \frac{-(-4) \pm \sqrt{(-4)^2 - 4(1)(7)}}{2(1)} = \frac{4 \pm \sqrt{-12}}{2}$$

$$x = \frac{4 \pm 2i\sqrt{3}}{2}$$

$$x = 2 + i\sqrt{3}, \; x = 2 - i\sqrt{3}$$

This equation has complex roots.

❹ $x^2 - 8x + 16 = 0$

$a = 1$, $b = (-8)$, $c = 16$

$b^2 - 4ac = (-8)^2 - 4(1)(16) = 0$

This equation has one root (also called a double root) and its graph is tangent to the x-axis. It can be factored.

$(x - 4)(x - 4) = 0$

$x - 4 = 0 \qquad x - 4 = 0$

$\qquad x = 4 \qquad\qquad x = 4$

Two equal roots are also described as one root or as a double root depending on the text.

Using the Discriminant with Graphs and with Equations

Use the rules for the discriminant to answer these questions.

Examples

❶ Sketch a graph to demonstrate when zero is the value of the discriminant of the equation $ax^2 + bx + c = 0$.

Solution:

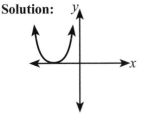

Rule: If $b^2 - 4ac = 0$ there is one real root. The graph is tangent to the x-axis.

❷ In the equation $ax^2 + 6x - 9 = 0$, imaginary roots will be generated for what values of a?

Rule: The roots are imaginary if $b^2 - 4ac < 0$

Solution:
$$6^2 - 4a(-9) < 0$$
$$36 + 36a < 0$$
$$36a < -36$$
$$a < -1$$

❸ Determine the value(s) of k if the roots of the equation $x^2 + kx + 3 = 0$ are real. Round to the nearest tenth.

Rule: If the roots are real, then $b^2 - 4ac \geq 0$. The only time they aren't real is when the discriminant is less than zero.

Solution:
$$b = k$$
$$k^2 - 4(1)(3) \geq 0$$
$$k^2 - 12 \geq 0$$
$$k^2 \geq 12$$
$$k \leq -\sqrt{12} \ \text{ or } k \geq \sqrt{12}$$
$$k \leq -3.5 \ \text{ or } k \geq 3.5$$

❹ What value of c would make the roots of the equation $x^2 + 6x + c = 0$ real, rational, and equal?

Rule: $b^2 - 4ac$ must equal zero for the roots to be real, rational, and equal. Both roots are the same – described as one root, or as a double root.

Solution:
$$6^2 - 4(1)c = 0$$
$$36 - 4c = 0$$
$$-4c = -36$$
$$c = 9$$

Arithmetic with Polynomials and Rational Expressions

THE SUM AND PRODUCT OF THE ROOTS OF A QUADRATIC EQUATION

The coefficients, a, b, and c of a quadratic equation in standard form, $ax^2 + bx + c = 0$, are closely related to the roots of the equation. We can find the roots using the coefficients, or we can find the coefficients and write an equation from the roots.

Formulas

Sum of the Roots: $r_1 + r_2 = \dfrac{-b}{a}$ **Product of the Roots:** $r_1 \bullet r_2 = \dfrac{c}{a}$

- a is assumed to be 1 unless there is information in the problem to indicate otherwise. If a is not 1, be sure to use its specified value in your calculations.

- To use the formulas, substitute the known information in the appropriate places and solve for the missing variable.

Examples

❶ Find the sum and product of the roots of the equation given:

$2x^2 + 3x - 4 = 0$

$a = 2, \quad b = 3, \quad c = -4$

Sum **Product**

$r_1 + r_2 = \dfrac{-b}{a}$ $\qquad$ $r_1 \bullet r_2 = \dfrac{c}{a}$

$r_1 + r_2 = \dfrac{-3}{2}$ $\qquad$ $r_1 \bullet r_2 = \dfrac{-4}{2} = -2$

Sum of roots: $-\dfrac{3}{2}$, **Product of roots:** -2

❷ The sum of the roots of a quadratic equation is 10 and the product is 9. Write an equation that has these roots.

Sum **Product**

$10 = \dfrac{-b}{1}$ $\qquad\qquad$ $9 = \dfrac{c}{1}$

$b = -10$ $\qquad\qquad\qquad$ $c = 9$

Solution: $x^2 - 10x + 9 = 0$

Note: Any equivalent equation will answer the question. Since we assume $a = 1$, this equation is the result of our calculations. An equivalent equation would also be correct, i.e. $2x^2 - 20x + 18 = 0$.

Algebra 2 Made Easy – Common Core Edition

❸ Write a quadratic equation with the roots equal to 1 and $\frac{1}{2}$ if $a = 2$.

$\left(1 + \frac{1}{2}\right) = \frac{-b}{2}$

$2\left(\frac{3}{2}\right) = -b$

$b = -3$

$(1)\left(\frac{1}{2}\right) = \frac{c}{2}$

$2\left(\frac{1}{2}\right) = c$

$c = 1$

Solution: $2x^2 - 3x + 1 = 0$

❹ One root of the equation $x^2 - 8x + c = 0$ is -3. Find the 2nd root and the value of c.   (*Hint:* Find the 2nd root first, then find c.)

$x^2 - 8x + c = 0$

$a = 1;\ b = -8;\ c = ?$

$r_1 + r_2 = \frac{-b}{a}$

$-3 + r_2 = \frac{-(-8)}{1}$

$-3 + r_2 = 8$

$r_2 = 11$

$r_1 \bullet r_2 = \frac{c}{a}$

$(-3)(11) = \frac{c}{1}$ (Substitute 11 for r_2)

$c = -33$ (Substitute -33 for
c in the equation)

Solution: The 2nd root is 11 and the equation is $x^2 - 8x - 33 = 0$

❺ Write an equation in which the roots are $2 + 3i$ and $2 - 3i$

$(2 + 3i) + (2 - 3i) = \frac{-b}{1}$

$4 = -b$

$b = -4$

$(2 + 3i)(2 - 3i) = \frac{c}{1}$

$4 - 9i^2 = c$

$4 - 9(-1) = c$

$c = 13$

Solution: $x^2 - 4x + 13 = 0$ or an equivalent equation.

Arithmetic with Polynomials and Rational Expressions

Examples

❶ A girl standing on top of the bleachers in a stadium catches a foul ball. She tosses it up vertically directly over her head to celebrate. The bleachers top row is 75 feet above the ground. The path the ball takes is given by the function $h(t) = -16t^2 + 64t + 350$ where t is the time in seconds after the ball is tossed. What is the height of the ball at its maximum height? How much time has elapsed when the ball reaches its maximum height? *Only an algebraic solution will be accepted.*

Solution: To solve algebraically, find the equation for the axis of symmetry of the graph. Since the maximum point will be on the axis of symmetry, use that value to find t in $h(t)$.

Axis of Symmetry: Value of $h(t)$ at $t = 2$

$$x = \frac{-b}{2a}$$ $$h(2) = -16(2)^2 + 64(2) + 35$$

$$x = \frac{-64}{2(-16)}$$ $$h(2) = -64 + 132 + 35$$

$$x = 2$$ $$h(2) = 103$$

Conclusion: Two seconds after being tossed vertically overhead, the ball will reach 103 feet.

❷ Determine all values of k for which the roots of the equation $x^2 + 5x + k = 0$ are complex. Explain your reasoning.

Solution: If the roots of a quadratic equation are complex, or imaginary, the discriminant must be less than 0. The discriminant is $b^2 - 4ac$ where b is the coefficient of the x term, a is the coefficient of the squared term, and c is a constant. In this example $a = 1$, $b = 5$, and $c = k$.

$$b^2 - 4ac < 0$$
$$(5)^2 - 4(1)k < 0$$
$$25 - 4k < 0$$
$$-4k < 25$$
$$k > \frac{-25}{4}$$
$$k > -6.25$$

Conclusion: When k is greater than -6.25, the roots of $x^2 + 5x + k = 0$ are complex.

$$\boxed{1.6}$$

❸ Solve $x^2 - 15 = -9x$ algebraically. Express the roots to the *nearest hundredth*. In what quadrant will the maximum or minimum value of graph of $x^2 - 15 = -9x$ be located? Explain your reasoning.

Solution: This equation is not factorable. Use the quadratic formula.

$$x^2 + 9x - 15 = 0$$
$$a = 1, \ b = 9, \ c = -15$$
$$x = \frac{-b \pm \sqrt{b^2 - 4ac}}{2a}$$
$$x = \frac{-9 \pm \sqrt{9^2 - 4(1)(-15)}}{2(1)}$$
$$x = \frac{-9 + \sqrt{141}}{2} \ and \ x = \frac{-9 - \sqrt{141}}{2}$$
$$x = 1.44 \ and \ x = -10.44$$

Conclusion: The roots, to the nearest hundredth, are 1.44 and –10.44. The coefficient of the x^2 term is positive, so the graph will have a minimum point. Because the positive root is very close to zero and the negative root is –10.44, the minimum point will be in the 3rd quadrant.

❹ Solve for x: $x^2 + 8x + 16 = \frac{7}{4} + 16$

Solution: Examine the equation and notice that it is set up like an equation being solved by using the completing the square technique. Use that to finish solving.

$$x^2 + 8x + 16 = \frac{7}{4} + 16$$
$$(x + 4)^2 = \frac{7}{4} + 16$$
$$(x + 4)^2 = \frac{7}{4} + \frac{64}{4} = \frac{71}{4}$$
$$x + 4 = \pm\sqrt{\frac{71}{4}}$$
$$x = -4 \pm \frac{\sqrt{71}}{2}$$

Conclusion: $x = -4 + \frac{\sqrt{71}}{2} \ and \ x = -4 - \frac{\sqrt{71}}{2}$

SOLVE BY GRAPHING

- Graph by hand or use the calculator. The roots are the x values where $y = 0$, which is where the graph crosses the x axis. If there is no intersection of the graph with the x axis, there are no real roots to the equation.

  It is often necessary to estimate the roots when the graph is created by hand. Read the x value of the intersection of the graph with the x axis as closely as possible.

  Using the calculator: Type the equation into y_1 and graph it. Use 2nd CALC 2:zero to find the roots. The process may need to be done twice for both roots. If there is only one root, the zero function won't work because the calculator requires a sign change for y to locate the "zero". Use the min/max function on the calculator instead of "zero" because the root will coincide with the vertex of the graph. The roots are found when $y = 0$ which is on the x axis.

Examples

❶ $f(x) = x^2 + 3x - 2$

This graph has two real roots. The roots can be estimated as $x = -3.6$ and $x = 0.6$ by looking at the graph. When using the calculator, the roots came up as approximately -3.561 and 0.561. The "discriminant" of this equation is positive and is not a perfect square. There are two real irrational roots.

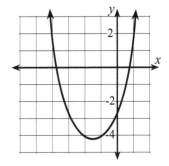

❷ $f(x) = x^2 - 6x + 9$

This equation has two equal roots, called a double root. Sometimes it is said to have one root. $x = 3$ is the root. The discriminant for this equation = 0.

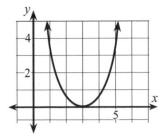

❸ $f(x) = x^2 - 2x + 3$

This equation has no real roots. It does not intersect with the x-axis. The discriminant for this equation is less than 0.

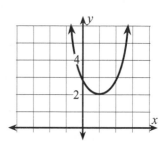

❹ The number of people, n, in thousands, who live in a town from 2010 to 2015 is approximated by the function $n(t) = 8.56t^2 + 121.5t + 5000$, where $t = 0$ corresponds to 2010. Use the model to determine in what year there will be 15,000 people living in the town.

Solution: Graph the function and use the table of values to determine when $n(t)$ reaches 15,000. The value of x at that point is 28. Since $t = 0$ corresponds to the year 2010, it is necessary to add 28 to 2010. In the year 2038, the population of the town will reach 15,000. (An alternate solution is to graph the function and graph $y = 15,000$ and find their point of intersection. It should be at (28, 15,000). Add 28 to 2010 to calculate the final answer of 2038.)

Conclusion: The population will reach 15,000 in 2038.

RATIONAL EXPRESSIONS

Rational Expression: A fraction with polynomials in the numerator and denominator is called a rational expression. It contains numbers and variables in various combinations of polynomials.

- A rational expression with a denominator = 0 is undefined. Therefore, we have to put restrictions on the variables to avoid a zero denominator.

Example $\dfrac{x+4}{x-3}$, $x \neq 3$ Because $x - 3 = 0$, x cannot be 3.

SIMPLIFYING RATIONAL EXPRESSIONS

Just as we can cancel factors in a rational number, we can cancel factors in a rational expression. (Divide the numerator and denominator by a common factor.)

Examples

❶ $\dfrac{12}{20} = \dfrac{3 \cdot \cancel{4}}{5 \cdot \cancel{4}} = \dfrac{3}{5}$, 4 is the common factor here.

❷   In this example, $(x + 3)$ is a common factor and x cannot be –3 or 5 since those values of x would cause the denominator of the original fraction to be zero.

$\dfrac{(x-2)\cancel{(x+3)}}{\cancel{(x+3)}(x-5)} = \dfrac{(x-2)}{(x-5)}$, $x \neq -3, 5$

Simplest Form: When a fraction has no factors in the numerator that are also in the denominator. Both examples above are in simplest form.

Steps:

1) Factor the numerator and denominator completely.

2) Cancel factors that are in both the numerator and the denominator.

3) Check to make sure no common factors remain.

4) Perform any indicated operations in the numerator and in the denominator.

(Examples on the next page)

Examples Simplest Form

① $\dfrac{12x^3}{12x^2 + 4x} = \dfrac{\cancel{4} \cdot 3 \cdot x\cancel{x}x}{\cancel{4}\cancel{x}(3x + 1)} = \dfrac{3x^2}{(3x + 1)}, \ x \neq 0, -\dfrac{1}{3}$

② $\dfrac{x^2 - 2x - 3}{x^2 - x - 2} = \dfrac{(x - 3)\cancel{(x + 1)}}{(x - 2)\cancel{(x + 1)}} = \dfrac{(x - 3)}{(x - 2)}, \ x \neq 2, -1$

③ $\dfrac{5x^2}{10x^3} = \dfrac{1}{2x}, \ x \neq 0$

④ $\dfrac{6x^2}{12x^2 - 6x} = \dfrac{\cancel{6}\cancel{x}\cancel{x}}{\cancel{6}\cancel{x}(2x - 1)} = \dfrac{x}{(2x - 1)}, \ x \neq 0, \dfrac{1}{2}$

⑤ $\dfrac{x^3 - 6x^2 + 9x}{x^2 - 9} = \dfrac{x(x^2 - 6x + 9)}{(x - 3)(x + 3)} = \dfrac{x\cancel{(x - 3)}(x - 3)}{\cancel{(x - 3)}(x + 3)} = \dfrac{x(x - 3)}{(x + 3)} = \dfrac{x^2 - 3x}{(x + 3)}, \ x \neq \pm 3$

Special Factors: When there are factors that are additive inverses (opposites) of each other but are not an exact match, we can create matching factors by factoring a (–1) out of one of them. This changes the signs of the resulting factor and we can reverse the order in the ( ) so it matches the factor in the opposite part of the fraction. For example, if a factor is $(4 - y)$ in the numerator and $(y - 4)$ in the denominator, use this method to change the numerator into $-1(y - 4)$: $(4 - y) = -1(-4 + y) = -1(y - 4)$

Examples Special Factors

① $\dfrac{(x - 5)(x + 2)}{(5 - x)(x - 1)} = \dfrac{(x - 5)(x + 2)}{-1(-5 + x)(x - 1)} = \dfrac{\cancel{(x - 5)}(x + 2)}{-1\cancel{(x - 5)}(x - 1)} = \dfrac{(x + 2)}{-(x - 1)} \ or \ -\dfrac{(x + 2)}{(x - 1)}$

② $\dfrac{(y - x)(x + 3)}{(x - 3)(x - y)} = \dfrac{-1(x - y)(x + 3)}{(x - 3)(x - y)} = -\dfrac{(x + 3)}{(x - 3)}$

OPERATIONS WITH RATIONAL EXPRESSIONS

Multiplying and Dividing Rational Expressions: The rules for multiplying and dividing algebraic fractions are the same as the rules for numeric fractions.

$$\frac{a}{b} \cdot \frac{c}{d} = \frac{ac}{bd} \quad and \quad \frac{a}{b} \div \frac{c}{d} = \frac{a}{b} \cdot \frac{d}{c} = \frac{ad}{bc}$$

- **Multiplication of Fractions:** Directions may say "perform the indicated operation" or "simplify" or "express the product in simplest form." Remember "product" indicates multiplication.

  Steps:

  1) If a monomial term, or a number, is by itself and is a factor in the problem, it is considered as a numerator. Put the monomial over "1" to make it look like a fraction if that makes it more clear. For example, $3x$ can be written as $\dfrac{3x}{1}$

  2) Factor all polynomials first. If a numerator or denominator is a binomial that cannot be factored, put parentheses around it.

  3) Cancel one factor in a numerator with a matching factor in a denominator until no common factors are left.

  4) Multiply the numerators together, multiply the denominators together.

  5) Review to make sure the fraction is in simplest form.

Examples

❶ $\quad \dfrac{1}{15x} \cdot 3x = \dfrac{1}{\underset{5}{\cancel{15x}}} \cdot \dfrac{\overset{1}{\cancel{3x}}}{1} = \dfrac{1 \cdot 1}{5 \cdot 1} = \dfrac{1}{5}$

❷ $\quad \dfrac{x+2}{3} \cdot \dfrac{5}{x} = \dfrac{5(x+2)}{3 \cdot x} = \dfrac{5(x+2)}{3x} \quad or \quad \dfrac{5x+10}{3x}$ Nothing to cancel.

❸ $\quad \dfrac{\overset{2x}{\cancel{10x^2}}}{\underset{y^2}{\cancel{8y^3}}} \cdot \dfrac{\overset{3}{\cancel{15x}}}{\underset{1}{\cancel{5x}}} = \dfrac{2x}{y^2} \cdot \dfrac{3}{1} = \dfrac{6x}{y^2}$

❹ $\quad \dfrac{4y^2-9}{4y^2} \cdot \dfrac{8y}{2y-3} = \dfrac{(2y+3)\cancel{(2y-3)}}{\underset{y}{\cancel{4y^2}}} \cdot \dfrac{\overset{2}{\cancel{8y}}}{\underset{1}{\cancel{(2y-3)}}} = \dfrac{2(2y+3)}{y} \quad or \quad \dfrac{4y+6}{y}$

❺ $\quad \dfrac{x^2-11x+24}{x^2-18x+80} \cdot \dfrac{x^2-15x+50}{x^2-9x+20} = \dfrac{\cancel{(x-8)}(x-3)}{\underset{1}{\cancel{(x-8)}\cancel{(x-10)}}} \cdot \dfrac{\overset{1}{\cancel{(x-5)}}\cancel{(x-10)}}{(x-4)\cancel{(x-5)}} = \dfrac{(x-3)}{(x-4)}$

Algebra 2 Made Easy – Common Core Edition

- **Division of Fractions:**
  Steps:
  1) Keep the first fraction (or if given a complex fraction, the entire numerator) as is.

  2) Change the division sign to multiplication.

  3) Multiply by the reciprocal of the divisor (or the denominator of the "double fraction") using the same method as always for multiplying fractions.

Examples

❶ $\dfrac{n^2 - 64}{4n} \div (8 - n) = \dfrac{n^2 - 64}{4n} \bullet \dfrac{1}{(8-n)} = \dfrac{\overset{1}{\cancel{(n-8)}}(n+8)}{4n} \bullet \dfrac{1}{-1\underset{1}{\cancel{(n-8)}}}$

$= \dfrac{n+8}{-4n}$ **or** $-\dfrac{n+8}{4n}$

❷ $\dfrac{\dfrac{10n^2 - 17n + 3}{5n^2 + 4n - 1}}{\dfrac{4n^2 - 9}{2n^2 + 5n + 3}} = \dfrac{10n^2 - 17n + 3}{5n^2 + 4n - 1} \bullet \dfrac{2n^2 + 5n + 3}{4n^2 - 9}$

$= \dfrac{\overset{1}{\cancel{(2n-3)}}\overset{}{\cancel{(5n-1)}}}{\cancel{(n+1)}\underset{1}{\cancel{(5n-1)}}} \bullet \dfrac{\overset{1}{\cancel{(n+1)}}\overset{}{\cancel{(2n+3)}}}{\cancel{(2n+3)}\underset{1}{\cancel{(2n-3)}}} = 1$

❸ $\dfrac{\dfrac{x^2 - 9}{2}}{x^2 - 2x - 3} = \dfrac{x^2 - 9}{2} \bullet \dfrac{1}{x^2 - 2x - 3} = \dfrac{\overset{1}{\cancel{(x-3)}}(x+3)}{2} \bullet \dfrac{1}{\underset{1}{\cancel{(x-3)}}(x+1)}$

$= \dfrac{x+3}{2(x+1)}$ **or** $\dfrac{x+3}{2x+2}$

Reasoning with Equations and Inequalities

Adding and Subtracting Rational Expressions: The least common denominator (LCD) is needed when adding or subtracting fractions, both algebraic and numerical. Each fraction in the problem must be made into an equivalent fraction with the LCD. Factoring the denominators helps us to determine what the LCD must contain.

Remember an LCD must be divisible by each denominator in the problem. It must contain all the factors that are in either denominator or in both denominators. If one denominator has two factors that are alike, the LCD must contain two of those factors. If there are two factors that are alike but they are in different denominators, then the LCD needs only one of those factors.

Steps:

1) Simplify each fraction if possible before beginning to factor.

2) Factor each denominator. Put a ( ) around each numerator.

3) Determine the factors needed for the LCD.

4) Write each fraction as an equivalent fraction with the LCD shown in factored form.

5) Make one new fraction – with a long fraction line – with the LCD. Rewrite the numerators, with their ( ) in the numerator. Make sure to keep the + or – sign between the numerators, separating them with brackets if necessary.

6) Use the distributive property in the numerator if needed and simplify by collecting like terms.

> *Note:* Steps 5 and 6 are equivalent to adding or subtracting the numerators of the fractions and keeping the LCD.

7) Make sure the final answer is in simplest form. Sometimes it is possible to factor the numerator after collecting like terms and then cancel a factor with a matching factor in the denominator.

Examples

❶ $\dfrac{x-5}{3} - \dfrac{4x+1}{2} = \dfrac{2(x-5)}{3 \cdot 2} - \dfrac{3(4x+1)}{2 \cdot 3} = \dfrac{2(x-5) - 3(4x+1)}{6}$

$= \dfrac{2x - 10 - 12x - 3}{6} = \dfrac{-10x - 13}{6}$

❷ Steps: Factor out –1   Factor denominators

$$\frac{x}{x-3} + \frac{6x}{9-x^2} = \frac{x}{(x-3)} + \frac{6x}{-1(x^2-9)} = \frac{x}{(x-3)} + \frac{6x}{-1(x-3)(x+3)}$$

Find LCD, make equivalent fractions   Make 1 fraction   Distribute –x

$$= \frac{[-1x(x+3)]}{-1(x-3)(x+3)} + \frac{6x}{-1(x-3)(x+3)} = \frac{[-1x(x+3)]+6x}{-(x-3)(x+3)} = \frac{-x^2-3x+6x}{-(x+3)(x-3)}$$

Combine terms   Factor   Cancel   Simplify

$$= \frac{-x^2+3x}{-(x+3)(x-3)} = \frac{-x\cancel{(x-3)}}{-(x+3)\cancel{(x-3)}} = \frac{-x}{-(x+3)} = \frac{x}{x+3}$$

COMPLEX FRACTIONS

Complex Rational Expression: A fraction that has a fraction in its numerator, denominator, or both. The numerator and/or denominator can contain "mixed" numbers – a number or variable and a fraction, or just fractions. These problems combine adding and subtracting, dividing and multiplying.

Steps:

1) Make the numerator into a single fraction and the denominator into a single fraction. Use the rules for adding and subtracting fractions to do this.

2) Now the problem becomes a fraction divided by a fraction, so the rules for division of fractions are needed. Keep the numerator fraction as it is and multiply it by the reciprocal of the denominator.

3) As always, simplify the answer whenever possible.

Examples

❶ $$\frac{\frac{x^2}{16}-1}{\frac{x}{8}-\frac{1}{2}} = \frac{\frac{x^2-16}{16}}{\frac{x-4}{8}} = \frac{x^2-16}{16} \cdot \frac{8}{x-4} = \frac{(x-4)(x+4)}{\overset{}{16}\;_2} \cdot \frac{\overset{1}{\cancel{8}}}{(x-4)} = \frac{x+4}{2}$$

❷ $$\frac{y+\frac{1}{2}}{y-\frac{1}{2}} = \frac{\frac{2y+1}{2}}{\frac{2y-1}{2}} = \frac{2y+1}{2} \cdot \frac{2}{2y-1} = \frac{2y+1}{2y-1}$$

❸ $$\frac{\frac{x}{4}-1}{\frac{x^2}{16}-1} = \frac{\frac{x-4}{4}}{\frac{x^2-16}{16}} = \frac{\overset{1}{(x-4)}}{\underset{1}{\cancel{4}}} \cdot \frac{\overset{4}{\cancel{16}}}{(x+4)(x-4)} = \frac{4}{x+4}$$

SOLVING RATIONAL EQUATIONS & INEQUALITIES

SOLVING RATIONAL EQUATIONS

Rational Equations: Equations that contain algebraic fractions (rational expressions). There are several methods available to solve them.

Proportions: Two rational expressions are equal to each other. "Cross multiply" is the common name for solving these.

Steps:

1) Multiply the numerator of each fraction by the denominator of the other.

2) Solve the resulting equation for the variable.

3) Checking is essential.

Note: Remember that some answers may cause a denominator in the problem to be equal to zero and must, therefore, be rejected. This will be evident when you check your answer(s) in the original fractions.

Example

$$\frac{x+4}{3} = \frac{4}{x}$$

$$x(x+4) = (3)(4)$$

$$x^2 + 4x = 12$$

$$x^2 + 4x - 12 = 0$$

$$(x+6)(x-2) = 0$$

$$x + 6 = 0 \quad x - 2 = 0$$

$$x = -6, \quad x = 2$$

Both answers are fine. Neither answer creates an undefined fraction.

If an equation is not a proportion, you can make it into one by changing each side of the equation into a single fraction, each side with its own LCD. Then cross multiply and solve.

Example

$$\frac{1}{9} + \frac{1}{2y} = \frac{1}{y^2}$$

$$\frac{2y+9}{18y} = \frac{1}{y^2}$$

$$(y^2)(2y+9) = (1)(18y)$$

$$2y^3 + 9y^2 = 18y$$

$$2y^3 + 9y^2 - 18y = 0$$

$$y(2y^2 + 9y - 18) = 0$$

$$y(y+6)(2y-3) = 0$$

$$y = 0 \quad y + 6 = 0 \quad 2y - 3 = 0$$

$$y = 0 \quad y = -6 \quad y = \frac{3}{2}$$

$y = 0$ has to be rejected. The answers are -6 and $\frac{3}{2}$

Algebra 2 Made Easy – Common Core Edition

$$\boxed{1.8}$$

Solve by Clearing the Fractions: This method removes the fractions from the problem.

Steps:

1) Find the Least Common Multiple (LCM) of ALL the denominators in the equation. Keep it in factor form.

2) Using the distributive property, multiply each term in the equation by the LCM, canceling whenever possible. This process removes the fractions completely.

3) Solve.

4) Check. Watch for zero denominators. All solutions may not check.

Examples

❶ $\dfrac{1}{9} + \dfrac{1}{2x} = \dfrac{1}{x^2}$ LCM is $18x^2$

$\overset{2}{\cancel{18x^2}}\left(\dfrac{1}{\cancel{9}}\right) + \overset{9x}{\cancel{18x^2}}\left(\dfrac{1}{\cancel{2x}}\right) = 18\cancel{x^2}\left(\dfrac{1}{\cancel{x^2}}\right)$

$2x^2 + 9x = 18$

$2x^2 + 9x - 18 = 0$

$(x + 6)(2x - 3) = 0$

$x + 6 = 0 \qquad 2x - 3 = 0$

$x = -6 \qquad\quad x = \dfrac{3}{2}$

❷ $\dfrac{3}{x} + \dfrac{2}{x + 2} = \dfrac{-x}{x + 2}$ LCM is $x(x + 2)$

$\cancel{x}(x + 2)\left(\dfrac{3}{\cancel{x}}\right) + x\cancel{(x + 2)}\left(\dfrac{2}{\cancel{x + 2}}\right) = x\cancel{(x + 2)}\left(\dfrac{-x}{\cancel{x + 2}}\right)$

$3(x + 2) + x(2) = (x)(-x)$

$3x + 6 + 2x = -x^2$

$x^2 + 5x + 6 = 0$

$(x + 2)(x + 3) = 0$

$x + 2 = 0 \qquad x + 3 = 0$

$x = -2 \qquad\quad x = -3$

$x = -2$ is rejected. $x = -3$ is the answer.

SOLVING RATIONAL INEQUALITIES

A **<u>rational inequality</u>** includes an inequality symbol instead of an equal sign.

Steps:

1) Determine what values of the variable will cause a fraction to be undefined.

2) Change the inequality symbol to = and solve the equation as usual.

3) On a number line, graph the solution(s) to the equation *and* the values of the variable found in step 1. These are called critical points.

4) The number line will be divided into sections. Test a number in each section of the number line to see which section(s) of the number line contain the numbers to make the inequality true.

5) Write the solution in inequality or set builder form.

Examples

❶ $\dfrac{3x - 4}{8} < \dfrac{4x - 3}{4}$

$4(3x - 4) = 8(4x - 3)$

$12x - 16 = 8(4x - 3)$

$12x - 16 = 32x - 24$

$-20x = -8$

$x = \dfrac{-8}{-20} = \dfrac{2}{5}$

Since this inequality has no variables in the denominator, only one point can be graphed. Test a value from each part of the number line in the original problem.

Test : 0 $\dfrac{3(0) - 4}{8} < \dfrac{4(0) - 3}{4}$

$-\dfrac{1}{2} < -\dfrac{3}{4}$ *False*

Test : 5 $\dfrac{3(5) - 4}{8} < \dfrac{4(5) - 3}{4}$

$\dfrac{11}{8} < \dfrac{17}{4}$ *True*

Solution: $\{x : x > \dfrac{2}{5}\}$

Algebra 2 Made Easy – Common Core Edition

SOLVING RADICAL EQUATIONS

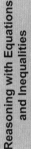

Radical Equations: Equations that contain radicals. To solve them, we must remove the radical.

Steps To solve:

1) Isolate the radical.
2) Square the entire equation on both sides, (or raise it to the nth power as indicated by the index).
3) Solve and CHECK - CHECKING IS ESSENTIAL for these problems. Extraneous roots are possible.

Examples

❶
$$7 + \sqrt{x - 3} = 1$$

$$\underline{-7 \qquad\qquad -7}$$

$$\left(\sqrt{x-3}\right)^2 = \left(-6\right)^2$$

$$x - 3 = 36$$

$$x = 39$$

Check

$$7 + \sqrt{x-3} \overset{?}{=} 1$$

$$7 + \sqrt{39-3} \overset{?}{=} 1$$

$$7 + \sqrt{36} \overset{?}{=} 1$$

$$7 + 6 \neq 1$$

Remember: $\sqrt{}$ indicates the principal root only.

Example 1 has no real solution. Although the work was done correctly, the answer does not check. This answer, $x = 39$, is called an extraneous solution.

❷
$$\sqrt{x + 5} = 3$$

$$\left(\sqrt{x+5}\right)^2 = 3^2$$

$$x + 5 = 9$$

$$x = 4$$

Check

$$\sqrt{4 + 5} = 3$$

$$\sqrt{9} = 3$$

$$3 = 3 \checkmark$$

❸
$$\sqrt{x^2 - 21} - x = -3$$

$$\underline{\qquad +x \qquad +x}$$

$$\sqrt{x^2 - 21} = x - 3$$

$$\left(\sqrt{x^2 - 21}\right)^2 = \left(x - 3\right)^2$$

$$x^2 - 21 = x^2 - 6x + 9$$

$$\underline{-x^2 - 9 \quad -x^2 \qquad -9}$$

$$-30 = -6x$$

$$x = 5$$

Check

$$\sqrt{5^2 - 21} - 5 = -3$$

$$\sqrt{4} - 5 = -3$$

$$2 - 5 = -3$$

$$-3 = -3 \checkmark$$

❹ $x - 2\sqrt{4x-3} = 3$

$$\frac{-x \qquad\qquad -x}{-2\sqrt{4x-3} = 3 - x}$$

$\left(-2\sqrt{4x-3}\right)^2 = \left(3-x\right)^2$

$4(4x-3) = 9 - 6x + x^2$

$16x - 12 = x^2 - 6x + 9$

$$\frac{-16x+12 \qquad\qquad -16x+12}{0 = x^2 - 22x + 21}$$

$\left(x-21\right)\left(x-1\right) = 0$

$x - 21 = 0;\ x - 1 = 0$

$x = 21, \qquad x = 1$

Check both answers

$21 - 2\sqrt{4(21)-3} = 3$

$21 - 2\sqrt{81} = 3;\ 21 - 18 = 3\ \checkmark$

$1 - 2\sqrt{4(1)-3} = 3$

$1 - 2\sqrt{1} = 3,\ 1 - 2 \neq 3$ *Doesn't check*

$x = 21$ *is the only correct answer.*

$x = 1$ *is an extraneous root.*

> For convenience, the coefficient (–2) is kept with the radical before squaring in this problem.

❺ $2\sqrt[3]{2x} = 10$

$\sqrt[3]{2x} = 5$

$\left(\sqrt[3]{2x}\right)^3 = \left(5\right)^3$

$2x = 125$

$x = 62.5$ **Check:** $2\sqrt[3]{2(62.5)} = 10;\ 2(\sqrt[3]{125}) = (2)(5) = 10\ \checkmark$

SYSTEMS OF EQUATIONS

What is a "system of equations"? When two or more equations are to be solved at the same time, it is called a system. The solution(s) of the system are the values of x and y at which the equations are equal to each other. When graphed, the solution(s) are the coordinates of the points of intersection of the graphs.

Linear Equation: A first degree equation which will form a straight line when its solution set is graphed. It can contain one or two variables.

Linear Pairs: Two first degree equations that are solved together. Linear pairs can be solved algebraically or graphically. Their solution set as a pair of equations is the ordered pair whose values make both equations true. Sometimes a linear pair has no solution.

Graphing: Each equation will be graphed on the same coordinate axis (same graph). The solution is the point(s) of intersection of the two graphs. (See also page 63.) If the lines are parallel there is no solution. If the two lines are concurrent (in the same place) the solution is all the real numbers.

Algebraic Solution: The algebraic solutions will check in both equations. Two methods can be used, substitution or addition.

Reasoning with Equations and Inequalities

SOLVE SYSTEMS ALGEBRAICALLY

Substitution: One equation is manipulated so that x or y is isolated, then the resulting representation of that variable is substituted in the second original equation. The remaining variable is solved for, then that answer is substituted in either original equation to find the second variable.
$x - y = 1$ and $x - 2y = 3$

Steps:

1) Solve the first equation $(x - y = 1)$ for x : $\qquad x = y + 1$

2) Get the second equation $(x - 2y = 3)$ $\qquad (y + 1) - 2y = 3$
and substitute $(y + 1)$ for x in it. $\qquad -y + 1 = 3$
Solve for y: $\qquad \boxed{-y = 2}$
$\qquad\qquad y = -2$

3) Go back to an original equation: $\qquad x - y = 1$

4) Substitute -2 for y: $\qquad x - (-2) = 1$

5) Solve for x: $\qquad x + 2 = 1$

6) Indicate both answers: $\qquad \boxed{x = -1 \text{ and } y = -2}$

7) Checking both answers $\qquad x - y = 1 \qquad x - 2y = 3$
in both original equations $\quad -1 - (-2) = 1 \quad -1 - 2(-2) = 3$
is the final step: $\qquad\quad -1 + 2 = 1 \qquad -1 + 4 = 3$
$\qquad\qquad\qquad\qquad\qquad 1 = 1 \qquad\qquad 3 = 3$

Note: This method is recommended ONLY when the coefficient of x or y is 1. Coefficients other than 1 result in fractional substitutions which must be "cleared" or they rapidly become unmanageable.

Addition: If two equations have the same variable with opposite coefficients, we can add the equations together and variable. Sometimes it is necessary to multiply one equation make equivalent equations that can be used in this method.

Steps:

1) Arrange both equations using algebraic methods so the variables are underneath each other in position.

2) The goal is to eliminate one variable by adding the two equations together. Examine the variables and their coefficients. Find the least common multiple of the coefficients of either both x's or both y's.

3) Multiply each equation by a positive or negative number so the coefficients of the variable chosen are equal and opposite in sign.

4) Add the two equations together. One variable will disappear.

5) Solve for the variable that is visible.

6) Choose one of the original equations and substitute the value for the known variable to find the other variable.

7) *Check* in *both original* equations.

Example Solve the following equations for x and y:

$$(A)\ 4x + 6y = 64 \qquad (B)\ 2x - 3y = -28$$

Steps:

1) 6 is a common multiple for 6 and 3 so leave (A) as is and multiply (B) by 2.

$$2(2x - 3y = -28)\ ;\ 4x - 6y = -56$$

2) Add (A) and (B) in order to isolate x:

Equation (A): $\qquad 4x + 6y = 64$

NEW Equation (B) from step 1: $\qquad 4x - 6y = -56$

$$8x = 8\ ;\ x = 1$$

3) Insert the answer back into either (A) or (B): $4(1) + 6y = 64$

4) Solve for y: $\qquad\qquad\qquad\qquad 6y = 60\ ;\ y = 10$

5) Check in both

original equations. $4(1) + 6(10) = 64 \qquad 2(1) - 3(10) = -28$

$$64 = 64\sqrt{} \qquad\qquad -28 = -28\sqrt{}$$

Note: Experience will help you decide which variable to work with in the addition method. Looking for variables which already have opposite signs allows you to avoid multiplying by a negative number with its associated opportunities for error. Using small multiples is helpful, too. You wouldn't want to use a common multiple for 11 and 14 if you could use the other variable and have a common multiple of 2 and 5.

Reasoning with Equations and Inequalities

1.10

WORD PROBLEM SYSTEMS WITH 2 VARIABLES

As in all word problems, read and reread. Make sure your equations represent the phrases in the wording of the problem.

1. Identify each unknown quantity and represent each one with a different variable in a let statement. READ CAREFULLY - make the let statement accurate.
2. Translate the verbal sentences into *two* equations.
3. Solve as a system of equations. Usually these problems are solved algebraically but follow directions - you might be directed to solve them graphically.
4. Check the answers in the words of the problem.

Word Problem Systems: Many word problems can be set up using two variables instead of using just one. If you choose to use that method, you must then make two equations to solve. The problem will then be solved as a "system of equations."

Example Together Evan and Denise have 28 books. If Denise has four more than Evan, how many books does each person have?

Let x = the number of books Evan has
Let y = the number of books Denise has

Steps:

1) Set up equation (A): $\qquad\qquad x + y = 28$

2) Set up equation (B): $\qquad\qquad y = x + 4$

3) Use substitution: $\qquad\quad x + (x + 4) = 28$

$$2x + 4 = 28$$
$$\underline{-4 \quad -4}$$

4) Solve for x: $\qquad\qquad\quad 2x = 24$

$$x = 12$$
$$x + y = 28$$

5) Substitute in original: $\qquad 12 + x = 28$

6) Solve for y: $\qquad\qquad\quad \underline{-12 \quad\; -12}$

$$y = 16$$

Answer: Evan has 12 books and Denise has 16 books.

SOLVE SYSTEMS BY GRAPHING

Graphing Systems of Linear Equations: Two or more equations are graphed on the same coordinate plane (grid).

- Graph EACH equation separately, but put both on one coordinate graph. Be ACCURATE.
- Label each line as you graph it.
- The point where they intersect is the *solution set* of the system of equations.
- Label the point of intersection on the graph. This point is the solution set.
- Check both the *x* and *y* values of the solution in both original equations. The *x* and *y* values of the point of intersection must satisfy both equations.

Example Solve this system of equations graphically and check:
(A) $y = x + 4$ and
(B) $y = -2x + 1$ (These are both already in $y = mx + b$ form.)

Steps: (A) $y = x + 4$ (B) $y = -2x + 1$

1) Determine the slope: slope = 1/1   slope = –2/1

2) Determine the *y*-intercept: (0, 4)   (0, 1)

3) Graph, label, and find solution set: SS = {(–1, 3)}

4) Check: (A) $y = x + 4$ (B) $y = -2x + 1$

                           $3 = -1 + 4$ $3 = -2(-1) + 1$

                           $3 = 3\checkmark$ $3 = 2 + 1 \Rightarrow 3 = 3\checkmark$

Reasoning with Equations and Inequalities

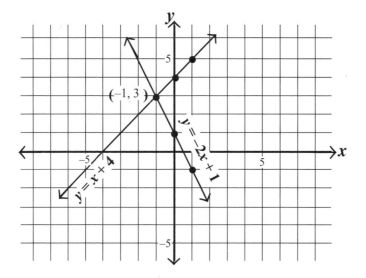

SOLVING SYSTEMS OF EQUATIONS WITH 3 VARIABLES

When problems with 3 variables are presented for solving, work with two equations at a time using substitution or addition. Each pair will result in a new equation. Solve the two new equations for one variable using addition or substitution. Substitute for the known variable to find a second variable. Then substitute both of the known variables in one of the original equations to find the third variable. Check.

Note: A system of 3 equations with 3 variables is not solved by graphing. In this course, our graphing is done on a two dimensional coordinate plane. Three variables would require a 3-dimensional graph.

Procedure: Solve these equations
for x, y, and z.

a) $2x - 3y + 2z = -2$
b) $x + 3y - z = 14$
c) $x + y - z = 10$

Steps:

1) Use equations **a)** and **b)**.
   Add or eliminate y in the
   the equation.

   a) $2x - 3y + 2z = -2$
   b) $\underline{x + 3y - z = 14}$
   $3x + z = 12$

2) Use equations **a)** and **c)**. Multiply equation **c)** by 3, then add the equations to eliminate y.
   a) $2x - 3y + 2z = -2$
   b) $x + y - z = 10 \implies 3(x + y - z = 10) \implies$

   $2x - 3y + 2z = -2$
   $\underline{3x + 3y - 3z = 30}$
   $5x - z = 28$

3) Use equations from steps **(1)** and **(2)**. Solve equation **(1)** for z. Substitute for z in equation **(2)**.
   Solve **(1)** for z: $\quad 3x + z = 12 \implies z = 12 - 3x$
   Substitute in **(2)** for z: $\quad 5x - z = 28 \implies 5x - (12 - 3x) = 28$
   Solve for x: $\qquad 5x - 12 + 3x = 28$
   $\qquad\qquad\qquad 8x - 12 = 28$
   $\qquad\qquad\qquad 8x = 40$
   $\qquad\qquad\qquad x = 5$

4) Use equation **(1)** and substitute for x to find z.

   (1) $3x + z = 12$
   $3(5) + z = 12$
   $z = -3$

5) Substitute the values of x and z in equation **a)**, **b)**, or **c)** to find the value of y.

   $x + y - z = 10$
   $5 + y - (-3) = 10$
   $y + 8 = 10$
   $\boxed{y = 2}$

Algebra 2 Made Easy – Common Core Edition

Check: The answers must check in all 3 original equations, **a)**, **b)**, and **c)**.

a) $2x - 3y + 2z = -2 \Rightarrow 2(5) - 3(2) + 2(-3) = 10 - 6 - 6 = -2$ √

b) $x + 3y - z = 14 \Rightarrow (5) + 3(2) - (-3) = 5 + 6 + 3 = 14$ √

c) $x + y + z = 10 \Rightarrow (5) + (2) - (-3) = 5 + 2 + 3 = 10$ √

Conclusion: $x = 5, y = 2, z = -3$

❶ Solve for x, y, and z: $a)$ $x + y + z = 9$
$b)$ $x - y - 2z = -17$
$c)$ $3x + 3y - z = 3$

Steps:

1) Use equations a and c:

$$x + y + z = 9$$
$$\underline{3x + 3y - z = 3}$$
$$4x + 4y = 12$$

2) Use equations a and b:

$x + y + z = 9 \Rightarrow 2(x + y + z = 9) \Rightarrow 2x + 2y + 2z = 18$
$x - y - 2z = -17 \qquad\qquad \underline{x - y - 2z = -17}$
$\qquad\qquad\qquad\qquad\qquad\qquad\qquad 3x + y = 1$

3) Use equations from steps 1 and 2 : $\qquad 4x + 4y = 12$

Solve for y : $3x + y = 1 \Rightarrow y = -3x + 1$

Substitute for y : $4x + 4(-3x + 1) = 12$
$$4x - 12x + 4 = 12$$
$$-8x = 8$$
$$x = -1$$

4) Substitute to find y and z : $\quad 4x + 4y = 12 \qquad x + y + z = 9$
$4(-1) + 4y = 12 \quad (-1) + 4 + z = 9$
$4y = 16 \qquad\qquad z = 6$
$y = 4$

The check is left to the student.

(Remember to check all answers in all 3 questions.)

Conclusion: $x = -1, \ y = 4, \ z = 6$.

Algebra 2 Made Easy – Common Core Edition
Copyright 2016 © Topical Review Book Company Inc. All rights reserved.

Reasoning with Equations and Inequalities

❷ Solve this system of equations for x, y, and z.

$$x - 2y + 3z = 4$$
$$2x + y + z = 13$$
$$-3x + 2y - 2z = -4$$

Solution: Follow the steps below:

1) Choose two equations at a time and eliminate a variable by addition or substitution.

$x - 2y + 3z = 4$ $\qquad\qquad$ $x - 2y + 3z = 4$

$2x + y + z = 13 \Rightarrow 2(2x + y + z = 13) \Rightarrow \underline{4x + 2y + 2z = 26}$

$\qquad\qquad\qquad\qquad\qquad\qquad\qquad\qquad 5x + 5z = 30$

2) Repeat step 1 with the other pair of equations.

$2x + y + z = 13 \Rightarrow -2(2x + y + z = 13) \Rightarrow -4x - 2y - 2z = -26$

$-3x + 2y - 2z = -4 \qquad\qquad\qquad\qquad\quad \underline{-3x + 2y - 2z = -4}$

$\qquad\qquad\qquad\qquad\qquad\qquad\qquad\qquad -7x - 4z = -30$

3) Combine the resulting two equations and solve as a system.

$5x + 5z = 30 \Rightarrow 4(5x + 5z = 30) \qquad \Rightarrow \quad 20x + 20z = 120$

$-7x - 4z = -30 \Rightarrow 5(-7x - 4z = -30) \Rightarrow \underline{-35x - 20z = -150}$

$\qquad\qquad\qquad\qquad\qquad\qquad\qquad\qquad\qquad -15x = -30$

$$\boxed{x = 2}$$

4) Substitute values for the variables in the original equations to find the values of the remaining variables.

$5x + 5z = 30 \qquad\qquad\qquad 2x + y + z = 13$

$5(2) + 5z = 30 \qquad\qquad\qquad 2(2) + y + 4 = 13$

$\qquad \boxed{z = 4} \qquad\qquad\qquad\qquad\qquad \boxed{y = 5}$

QUADRATIC LINEAR SYSTEMS

To Solve Graphically
Steps:
1) Graph each equation – either by making a table of values and graphing or by using the graphing calculator. Sketch the graphs on graph paper and label appropriately. Label the parabola at the vertex and at one point on each side of the vertex. Label the line with two points. Both are labeled with their equations or $f(x)$ and $g(x)$ if written that way in the problem. Make sure the graphs extend to the edges of the graph paper.

2) Locate the point(s) of intersection of the graphs and label them with their coordinates. Approximate the points by inspecting the graph if you made it from "scratch" or by using the $\boxed{\text{2nd}}$ CALC *5:intersect* function of the calculator. **

3) State the solution set in the form of coordinates.
   Solution Set = $\{ (x_1, y_1), (x_2, y_2) \}$ Check in both original equations.

4) If the graphs do not intersect, there is no solution.
   Solution Set = $\{ \ \}$ *or* Solution Set = $\varnothing$

** Substitute the answers in both original equations to check for accuracy.

Solve Algebraically – Use Substitution.
(If the equations are written in function form, change $f(x)$ to y in both equations.)

Steps:
1) Solve the linear equation for y.

2) Substitute the resulting expression for y in the quadratic equation. This removes y from the quadratic equation and we can then solve for x.

3) Write the equation in standard form and make it equal to zero.

4) Solve as usual – factor, quadratic formula, etc. Find the real values for x.
   a) Two real answers for x means the equations are equal for two values of x and y. (The graphs would intersect in two places.)

Reasoning with Equations and Inequalities

    b)   One real answer indicates equations are equal at only one value of x and y. (The linear graph is tangent to the parabola.)

    c)   Complex or imaginary answers indicate "no real solution." (The graphs would not intersect.)

5)   Go back to the ORIGINAL equation(s) and substitute each value of x to find the corresponding value of y. (Although either equation can be used, the linear equation is usually easier to work with.)

6)   Check all sets of answers in BOTH ORIGINAL equations.

7)   Write the answers as coordinates or as pairs of solutions. Be careful to match them up correctly. $x = 5, y = 12$; $x = 3, y = 6$. *or* $\{(5, 12), (3, 6)\}$

Show your work neatly and be well organized when solving these multi-step problems algebraically. The reader of your paper must be able to follow the steps and to identify the answers.

Examples (Graphs are for demonstration and are not part of the algebraic solution.)

❶ $g(x) = x^2 - 3x + 5:$ $y = x^2 - 3x + 5$

    $s(x) = x + 2:$ $y = x + 2$

    Substitute : $x + 2 = x^2 - 3x + 5$

    Standard form = 0: $x^2 - 4x + 3 = 0$

    Solve : $(x - 3)(x - 1) = 0$

            $x = 3$ $x = 1$

    $y = x + 2$ $y = x + 2$

    $y = 3 + 2 = 5$ $y = 1 + 2 = 3$

    $x = 3, y = 5$ $x = 1, y = 3$

    Solution Set $= \{(3, 5), (1, 3)\}$

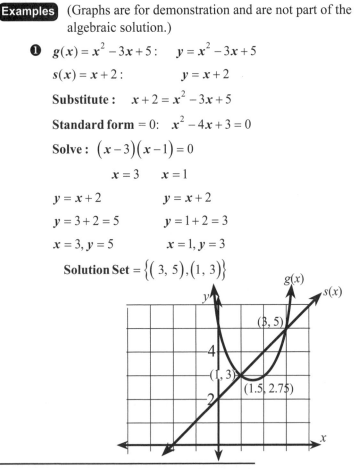

❷ $h(x) = 3x^2 - 6x - 2$: $y = 3x^2 - 6x - 2$

$g(x) = 2x - 5$: $y = 2x - 5$

$2x - 5 = 3x^2 - 6x - 2$

$3x^2 - 8x + 3 = 0$ *Use quadratic formula*

$$x = \frac{-(-8) \pm \sqrt{(-8)^2 - 4(3)(3)}}{2(3)}$$

$$x = \frac{8 \pm \sqrt{64 - 36}}{6}$$

$$x = \frac{8 \pm \sqrt{28}}{6} = \frac{\overset{4}{\cancel{8}} \pm \overset{1}{\cancel{2}}\sqrt{7}}{\cancel{6}\,3}$$

$$x = \frac{4 + \sqrt{7}}{3} \qquad\qquad x = \frac{4 - \sqrt{7}}{3}$$

$$y = 2\left(\frac{4 + \sqrt{7}}{3}\right) - 5 \qquad y = 2\left(\frac{4 - \sqrt{7}}{3}\right) - 5$$

$$y = \frac{8 + 2\sqrt{7}}{3} - \frac{15}{3} \qquad y = \frac{8 - 2\sqrt{7}}{3} - \frac{15}{3}$$

$$y = \frac{-7 + 2\sqrt{7}}{3} \qquad\qquad y = \frac{-7 - 2\sqrt{7}}{3}$$

$$\left\{ \left(\frac{4 + \sqrt{7}}{3}, \frac{-7 + 2\sqrt{7}}{3}\right), \left(\frac{4 - \sqrt{7}}{3}, \frac{-7 - 2\sqrt{7}}{3}\right) \right\}$$

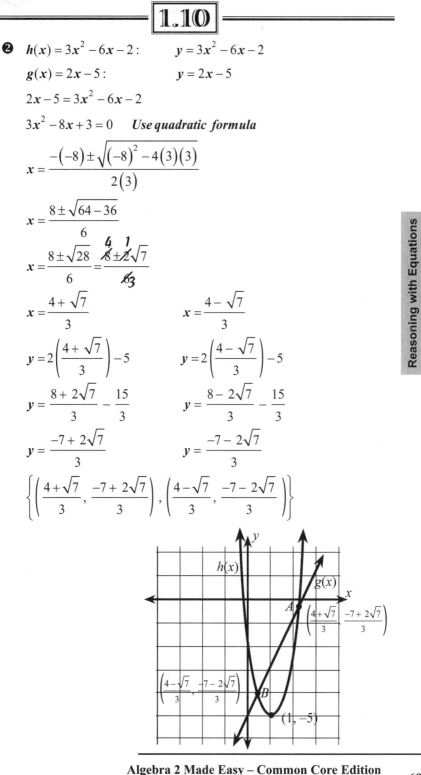

Reasoning with Equations and Inequalities

LINEAR EQUATIONS AND CIRCLES

Steps:

1) Solve the linear equation for y.

2) Substitute it in the circle equation in place of y.

3) Solve for x.

4) Then solve for y.

Example

$$x^2 + y^2 = 25$$

$$y = x + 1$$

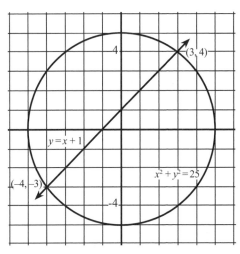

Steps:

1) $x^2 + (x+1)^2 = 25$

2) $x^2 + (x^2 + 2x + 1) - 25 = 0$

3) $2x^2 + 2x - 24 = 0$

4) $2(x^2 + x - 12) = 0$

5) $(x+4)(x-3) = 0$

6) $x = -4$ $\qquad\qquad$ $x = 3$

7) $y = x + 1$ $\qquad\qquad$ $y = x + 1$

8) $y = -4 + 1 = -3$ $\qquad$ $y = 3 + 1 = 4$

$\qquad$ $(-4, -3)$ $\qquad\qquad\qquad$ $(3, 4)$

$\qquad$ **Solution Set** $= \{(-4, -3), (3, 4)\}$

If a graph is given and you are asked to solve using the graph, make your best estimate of the intersection of the points. Make sure and check both sets of (x, y) in both equations. Show the checks on your paper.

Note: In a system of equations, solutions can be found by graphing all of the equations on the same graph. Points of common intersection are the solutions. If the system is three equations, the point(s) where all three intersect are the solutions. CHECK as usual.

PRACTICE AND SOLVING EQUATIONS

Examples

❶ Find the values(s) of x that solve $f(x) = g(x)$ algebraically. Explain your reasoning. Does this give any indication of what occurs if these two functions are graphed? Explain how to find the coordinates of any point(s) of intersection. $f(x) = |x + 6| + 1$

$$g(x) = \frac{1}{2}x + 4$$

Steps:

1) Set the equations equal to each other and simplify to isolate the absolute value term.

$$|x + 6| + 1 = \frac{1}{2}x + 6$$

$$|x + 6| = \frac{1}{2}x + 5$$

2) The expression inside the absolute value symbol can be positive or negative. Both possibilities must be considered. Remove the absolute value symbol and make $x + 6$ equal to the positive and the negative values of the right side of the equation. Solve both.

$$x + 6 = \frac{1}{2}x + 5 \qquad\qquad x + 6 = -\left(\frac{1}{2}x + 5\right)$$

$$\frac{1}{2}x = -1 \qquad\qquad\qquad x + 6 = -\frac{1}{2}x - 5$$

$$\boxed{x = -2} \qquad\qquad\qquad \frac{3}{2}x = -11$$

$$\boxed{x = \frac{-22}{3}}$$

3) Check both answers in the equation $f(x) = g(x)$.

$$|x + 6| + 1 = \frac{1}{2}x + 6 \qquad\qquad |x + 6| + 1 = \frac{1}{2}x + 6$$

$$|-2 + 6| + 1 = \frac{1}{2}(-2) + 6 \qquad |\frac{-22}{3} + 6| + 1 = \frac{1}{2}\left(\frac{-22}{3}\right) + 6$$

$$|4| + 1 = -1 + 6 \qquad\qquad |\frac{-4}{3}| + 1 = \left(\frac{-11}{3}\right) + 6$$

$$|4| = 4 \qquad\qquad\qquad\qquad \frac{4}{3} + 1 = \frac{-11}{3} + 6$$

$$4 = 4 \checkmark \qquad\qquad\qquad\qquad \frac{7}{3} = \frac{7}{3} \checkmark$$

Conclusion: $x = -2$ and $x = -22/3$ are the values that make $f(x) = g(x)$. Two answers for x that check indicate that the graphs of $f(x)$ and $g(x)$ will intersect at 2 points. To find the y values of the points of intersection, substitute the x values and evaluate $f(x)$ and $g(x)$ or use the intersection function on the graphing calculator. The x values of the points of intersection are $x = -2$ and $x = -22/3$. The coordinates of the point of intersection are $(-6, 1)$.

Solving Equations

❷ Graph the following equations and determine the approximate value(s) that result in an equal output for both functions. Why is the word "approximate" used in this question? How would you prove your answer is correct?

$$f(x) = 4^x$$
$$g(x) = -5x + 9$$

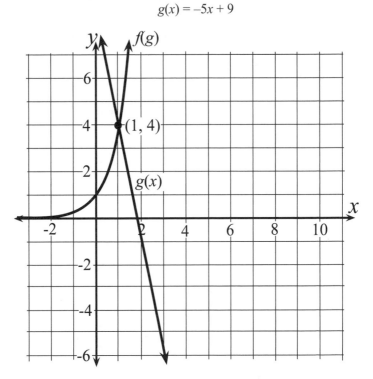

Conclusion: The functions are equal when $x = 1$ and $y = 4$. Approximate is used because a graph is a sketch of the two functions and it is not possible to read an exact value from the graph. To prove the answer is correct, find the values of $f(1)$ and $g(1)$ to determine if both are equal to 4.

Examples

❶ Kim's horse, Captain, is sick and the vet gave Kim the following information about giving Captain his medicine. The amount of the medicine in Captain's bloodstream is modeled by $f(x) = \dfrac{20}{t^2 + 1}$ where t is in hours and $f(x)$ is in milligrams per liter. Kim cannot give the horse a second dose until the medicine measures 3 milligrams per liter in his bloodstream. What is the shortest time that will elapse before Captain can be given a second dose?

Solution: Since $f(x)$ is the number of milligrams of medicine per liter, and 3 is the maximum amount of medicine that can be in the bloodstream before a second dose is given, substitute 3 for $f(x)$ and solve.

$$f(x) = \frac{20}{t^2 + 1}$$

$$3 = \frac{20}{t^2 + 1}$$

$$3(t^2 + 1) = 20$$

$$3t^2 + 3 = 20$$

Conclusion: Since time cannot be a negative value, the shortest time before the second dose can be given is approximately 2.38 hours.

$$3t^2 = 17$$

$$t = \pm\sqrt{\frac{17}{3}}$$

$$t \approx \pm 2.38$$

❷ A car is traveling on the highway and an accident occurs 200 feet ahead of the car. The stopping distance of the car on dry pavement is modeled by $d = 1.1v + 0.06v^2$ where d is the stopping distance in feet and v is the speed of the car in mph. What is the maximum speed the car can be traveling to avoid being involved in the accident?

Solution: The stopping distance must be less than 200 feet for the car to avoid the accident. Substitute 200 for d and solve. Any speed less than the value obtained for v will be the answer.

Conclusion: Speed cannot be a negative number so –67.62 is rejected. The car must be traveling less than about 49.29 mph.

$$d = 1.1v + 0.06v^2$$

$$200 = 1.1v + 0.06v^2$$

$$0.06v^2 + 1.1v - 200 = 0$$

$$a = 0.06, \ b = 1.1, \ c = -200$$

$$v = \frac{1.1 \pm \sqrt{(1.1)^2 - 4(0.06)(-200)}}{2(0.06)}$$

$$v = 49.29 \ \ or \ \ v = -67.62$$

❸ A ball is hit from a height of 5 feet with an initial velocity, v_0, of 85 feet per second. The equation $s = -16t^2 + v_0 t + s_0$ models the height of the ball, s, as a function, t. What is the height of the ball 5 seconds after being hit?

Solving Equations

Algebra 2 Made Easy – Common Core Edition

Solution: Since the ball was hit from a height of 5 feet, $s_0 = 5$. Initial velocity is $v_0 = 85$. The time in seconds is $t = 5$. Substitute those values in the equation and evaluate to find the value of s.

Conclusion: Five seconds after it was hit, the ball was at a height of 30 feet.

$$s = -16(5)^2 + 85(5) + 5$$
$$s = 30 \text{ feet}$$

❹ Takeya paints porcelain dolls to sell. First she has to purchase the dolls and the paint. It takes her 45 minutes to go to the store. It takes her 10 minutes to plan the design and choose the colors for each doll that she paints, and 20 minutes for her to paint each one. She has 4 hours to work on the project. Write an absolute value function, $d(x)$ to model the difference between the time she has available in minutes and the time it takes Takeya to complete the project making x dolls. What is the maximum number of dolls she can complete it 4 hours?

Solution: The 45 minutes she needs to go to the store is a one time occurrence. It takes a total of 30 minutes to complete each doll. To determine the maximum number of dolls she can complete, make an inequality where the 4 hours she has available is greater than or equal to the 45 minute shopping time and the 30 minutes per doll it takes to paint them.

Conclusion: The model for the difference in time is $d(x) = 4(60) - |45 + 30x|$. She can pain 6 dolls in the time she has available.

$$d(x) = 4(60) - |45 + 30x|$$
$$4(60) \geq 45 + 30x$$
$$240 \geq 45 + 30x$$
$$195 \geq 30x$$
$$x \leq 6.5$$

❺ Allan and Jose are working together to unload a truck of supplies where they work. It would take 6 hours for Allan to do the job by himself and Jose works one and one-half times as fast as Allan. How long does it take them to unload the truck if they work together?

Solution: Each person can do part of the work in one hour. Since Jose works 1.5 times as fast as Allan, he could do the job in 4 hours working alone. Add the two one hour values to find out how much the can do together in one hour.

Let t = time needed to complete job together.

Conclusion: Allan and Jose can completely unload the truck in 2.4 hours if they work together.

$$\frac{1}{6} + \left(\frac{1}{4}\right) = \frac{1}{t}$$
$$\frac{2}{12} + \frac{3}{12} = \frac{1}{t}$$
$$\frac{5}{12} = \frac{1}{t}$$
$$t = \frac{12}{5} = 2.4 \text{ hours}$$

PARABOLA: FOCUS AND DIRECTRIX

A quadratic equation forms a parabola when it is graphed.

Focus and Directrix: The focus is a point inside the curve of a parabola. The parabola curved away from a line is called the directrix. The vertex of the parabola is the point on the parabola that is half the distance between the focus and the directrix along a line perpendicular to the directrix. The distance from the focus to any point on the parabola is equal to the distance between that point on the parabola and a line drawn perpendicular to the directrix.

Figure 1

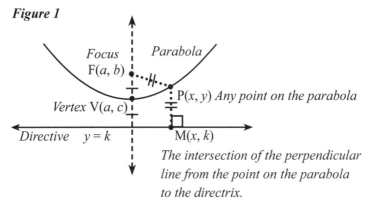

Focus F(a, b) Parabola P(x, y) *Any point on the parabola*

Vertex V(a, c)

Directive $y = k$ M(x, k)

The intersection of the perpendicular line from the point on the parabola to the directrix.

Expressing Geometric Properties with Equations

Equation of a Parabola: Two forms of a quadratic equation that will produce a parabola are:

Vertex Form: $y = r(x - a)^2 + c$, where the coordinates of the vertex are (a, c); and r is the coefficient of x^2. This labeling is shown on Figure 1. This equation is derived on the next page.

Standard Form: $y = ax^2 + bx + c$, where a is the coefficient of x^2, the squared term, b is the coefficient of x and c is a constant. This form is usually used when the quadratic formula is involved.

Note: The names of these equations and the variables used are different in various resources.

DERIVE THE EQUATION OF A PARABOLA
GIVEN A FOCUS AND DIRECTRIX

Since any point on the parabola is equidistant from the focus and the directrix along a line drawn perpendicular to the directrix, the distance formula can be used to derive an equation of a parabola. The distance between a point on the parabola to the focus, and the distance between the point on the parabola to the directrix, can be set equal to each other and the equation can be developed.

In Figure 2 a vertical parabola is given and $FP = PM$. In this case, PM is vertical, so its length is the absolute value of $y - k$.

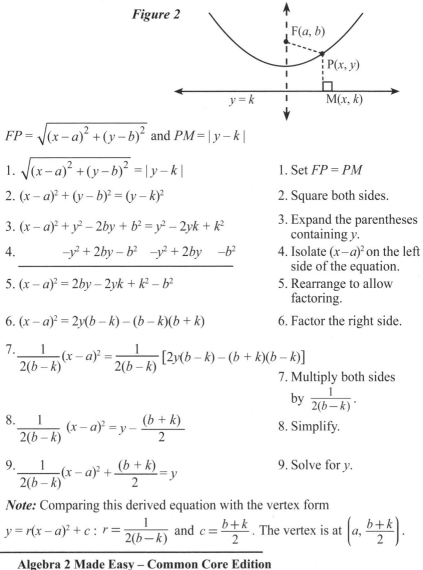

Figure 2

$FP = \sqrt{(x-a)^2 + (y-b)^2}$ and $PM = |y-k|$

1. $\sqrt{(x-a)^2 + (y-b)^2} = |y-k|$ 1. Set $FP = PM$

2. $(x-a)^2 + (y-b)^2 = (y-k)^2$ 2. Square both sides.

3. $(x-a)^2 + y^2 - 2by + b^2 = y^2 - 2yk + k^2$ 3. Expand the parentheses containing y.

4. $\dfrac{\quad -y^2 + 2by - b^2 \quad -y^2 + 2by \quad -b^2 \quad}{}$ 4. Isolate $(x-a)^2$ on the left side of the equation.

5. $(x-a)^2 = 2by - 2yk + k^2 - b^2$ 5. Rearrange to allow factoring.

6. $(x-a)^2 = 2y(b-k) - (b-k)(b+k)$ 6. Factor the right side.

7. $\dfrac{1}{2(b-k)}(x-a)^2 = \dfrac{1}{2(b-k)}\left[2y(b-k) - (b+k)(b-k)\right]$ 7. Multiply both sides by $\dfrac{1}{2(b-k)}$.

8. $\dfrac{1}{2(b-k)}(x-a)^2 = y - \dfrac{(b+k)}{2}$ 8. Simplify.

9. $\dfrac{1}{2(b-k)}(x-a)^2 + \dfrac{(b+k)}{2} = y$ 9. Solve for y.

Note: Comparing this derived equation with the vertex form

$y = r(x-a)^2 + c : r = \dfrac{1}{2(b-k)}$ and $c = \dfrac{b+k}{2}$. The vertex is at $\left(a, \dfrac{b+k}{2}\right)$.

Algebra 2 Made Easy – Common Core Edition

Examples

❶ Find the equation of a parabola with the focus at $F(2, 1)$ and the directrix at $y = -2$.

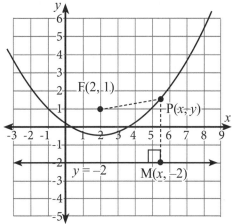

$F(2, 1)$ is the focus. The directrix is $y = -2$.

Draw a line from $P(x, y)$ perpendicular to the directrix $(y = -2)$ at point $M(x, -2)$.

The definition of a parabola tells us that $PM = FP$.
- Point P is y units above the x-axis and M is 2 units below the x-axis. The length of $PM = |y + 2|$.
- FP requires the use of the distance formula. $FP = \sqrt{(x-2)^2 + (y-1)^2}$

1. $y + 2 = \sqrt{(x-2)^2 + (y-1)^2}$

1. Set the lengths of PM and $FP =$ to each other.

2. $(y + 2)^2 = (x-2)^2 + (y-1)^2$

2. Square both sides. Square the binomial containing y as indicated.

3. $y^2 + 4y + 4 = (x-2)^2 + y^2 - 2y + 1$

$\underline{-y^2 + 2y - 4 -y^2 + 2y - 4}$

$ 6y = (x-2)^2 - 3$

3. Rearrange putting y on one side, x on the other.

4. $y = \dfrac{(x-2)^2 - 3}{6}$

4. Solve for y by dividing both sides by 6.

5. $y = \dfrac{1}{6}(x-2)^2 - \dfrac{1}{2}$

5. This equation is in vertex form. Vertex is at $(2, -1/2)$.

Expressing Geometric Properties with Equations

❷ The science class was split into 2 teams to work on a robot project. The whole class designed a robot that can follow an equation programmed into it along the path indicated by the equation. One team of students designed the course as shown on the diagram below. The other team will program the robot to travel the course. The robot must start at the back of the classroom and travel around the cone and return to the back of the room. At all times the robot must be equidistant from the cone and the front wall.

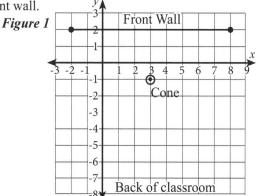

Figure 1

The team that is developing the correct equation determined that the path for the robot is a parabola. They put the diagram on a grid and labeled it. The class has worked on the equation of a parabola in their math class, and they have derived a formula that they know they can use. They label the cone as the focus at $F(3, -1)$ and the front wall line as the directrix at $y = 2$. Figure 2 shows the path the robot took to accomplish the task.

Finding the equation: Substitute $(3, -1)$ for a and b, and $k = 2$ since the directrix is at $y = 2$.

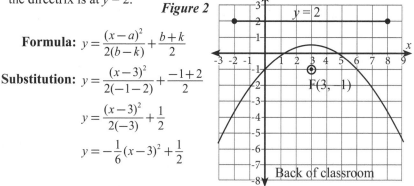

Figure 2

Formula: $y = \dfrac{(x-a)^2}{2(b-k)} + \dfrac{b+k}{2}$

Substitution: $y = \dfrac{(x-3)^2}{2(-1-2)} + \dfrac{-1+2}{2}$

$$y = \dfrac{(x-3)^2}{2(-3)} + \dfrac{1}{2}$$

$$y = -\dfrac{1}{6}(x-3)^2 + \dfrac{1}{2}$$

Programming the robot with the equation $y = -\dfrac{1}{6}(x-3)^2 + \dfrac{1}{2}$ allowed it to complete the path as the rules directed.

Algebra 2 Made Easy – Common Core Edition

Unit 2

TRIGONOMETRY FUNCTIONS

- Extend the domain of trigonometric functions using the unit circle.

- Model periodic phenomena with trigonometric functions.

- Prove and apply trigonometric identities.

- Summarize, represent and interpret data on two categorical and quantitative variables.

2.1

To find the length of the arc subtended (cut off by) by a central angle of a circle use this formula: $s = r\theta$ where s represents the arc length, r is the radius, and θ is the central angle MEASURED IN RADIANS. If the measure of the central angle is given in degrees, it must be changed to radian measure before using the formula.

Examples

❶ Find the length of the arc intercepted by θ if the radius is 10 cm and $\theta = 2$. Since θ is already in radian form, we don't need to change it.

$$s = r\theta$$
$$s = (10)(2)$$
$$s = 20 \text{ cm}$$

❷ Find θ in radians if the arc length is 12 and the radius is 4.

$$s = r\theta$$
$$12 = 4\theta$$
$$\theta = 3 \text{ radians}$$

❸ Find the length of the arc, to the nearest tenth, cut off by a central angle measuring $75°$ if the radius of the circle is 5 cm. *Hint: Change degrees to radians.*

$$75° \bullet \frac{\pi \text{ radians}}{180°} = \frac{75\pi}{180} = \frac{5\pi}{12} \text{ radians}$$

$$s = \left(\frac{5\pi}{12}\right)(5) = 6.54$$

$$s \approx 6.5 cm$$

---

❹ Find the arc length, to the nearest hundredth, intercepted by an angle, θ, that measures 110° in a circle with its center at (1, 3). A point on the circle is (7, 9).

Hint: Find the length of the radius using the distance formula and change degrees to radians!

$$d = \sqrt{(1-7)^2 + (3-9)^2} = \sqrt{72} = 6\sqrt{2} \quad (d = radius)$$

$$\theta = 110° \cdot \frac{\pi \text{ radians}}{180°} = \frac{11\pi}{18} \text{ radians}$$

$$s = \frac{11\pi}{18} \cdot 6\sqrt{2} \approx 16.29$$

❺ Find the length of the diameter of a circle, to the nearest hundredth, if a central angle, $\theta = \frac{5\pi}{7}$ radians, subtends an arc of 25 cm. Remember that the diameter = 2r.

$$25 = \frac{5\pi}{7} \cdot r$$

$$r = \frac{7}{5\pi} \cdot 25$$

$$r = 11.1408$$

$$d = 2r$$

$$d = 2(11.1408)$$

$$d = 22.2816; \quad d \approx 22.28 \text{ cm}$$

RADIANS AND DEGREES OF SPECIAL ANGLES

Special Angles
in radian measure

30°	$\frac{\pi}{180°} = \frac{30\pi}{180} = \frac{\pi}{6}$
45°	$\frac{\pi}{180°} = \frac{45\pi}{180} = \frac{\pi}{4}$
60°	$\frac{\pi}{180°} = \frac{60\pi}{180} = \frac{\pi}{3}$

Note: These angles are reference angles in Quadrants 2, 3, and 4.

Quadrantal Angles
in radian measure

0°	$\frac{\pi}{180°} = 0$
90°	$\frac{\pi}{180°} = \frac{\pi}{2}$
180°	$\frac{\pi}{180°} = \pi$
270°	$\frac{\pi}{180°} = \frac{3\pi}{2}$
360°	$\frac{\pi}{180°} = 2\pi$

Trigonometric Functions

ANGLES IN QUADRANTS

An angle in standard position has trig functions that are equal to the positive or negative value of the trig function of its reference angle. The reference angle is found by drawing a line perpendicular to the nearest x-axis from a point on the terminal side of the angle. This creates a right triangle. The angle formed by the terminal side of the angle and the x-axis is the reference angle. The (x, y) values of the point on the terminal side provide the lengths of the legs of the triangle, and the hypotenuse is determined using the Pythagorean Theorem. The coordinates (x, y) of the point on the terminal side can be positive or negative. The sign of the x or y is maintained with the length for use in the trig functions. The hypotenuse is always positive.

In the following diagrams, the main angle is named θ and the reference angle is named α.

Quadrant I: Since the nearest x-axis is the initial side of θ, the reference angle in Quadrant I is θ itself. Point A is on the terminal side of θ.

$0° < \theta < 90°$

θ *and* α *are* = .

$\sin \theta = \dfrac{y}{r}$

$\cos \theta = \dfrac{x}{r}$

$\tan \theta = \dfrac{y}{x}$

All **trig functions are positive in Quadrant I.**

Quadrant II: The nearest x-axis is the negative side of the x-axis. Point A is on the terminal side of θ. The coordinates of A are $(-x, y)$.

$90° < \theta < 180°$

$\alpha = 180 - \theta$

$\sin \alpha = \dfrac{y}{r}; \quad \therefore \sin \theta = \dfrac{y}{r}$

$\cos \alpha = \dfrac{-x}{r}; \quad \therefore \cos \theta = \dfrac{-x}{r}$

$\tan \alpha = \dfrac{y}{-x}; \quad \therefore \tan \theta = \dfrac{y}{-x}$

Only **Sin is positive in Quadrant II.**

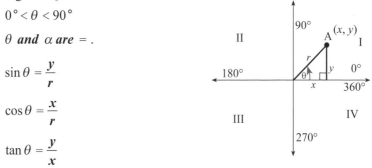

In Quadrant II: To find α, subtract θ from $180°$.

Quadrant III: The terminal side of θ is below the negative x-axis. The perpendicular line from A, on the terminal side, to the nearest x-axis goes up. Now x and y are both negative.

$180° < \theta < 270°$

$\alpha = \theta - 180°$

$\sin \alpha = \dfrac{-y}{r}$; ∴ $\sin \theta = \dfrac{-y}{r}$

$\cos \alpha = \dfrac{-x}{r}$; ∴ $\cos \theta = \dfrac{-x}{r}$

$\tan \alpha = \dfrac{-y}{-x} = \dfrac{y}{x}$; ∴ $\tan \theta = \dfrac{y}{x}$

***Only* Tan is positive in Quadrant III.**

In Quadrant III: To find α, subtract 180° from θ.

Quadrant IV: The positive x-axis is the closest to the terminal side of θ. Use the 360° value of the positive x-axis to determine the reference angle. In Q IV, x is positive and y is negative.

$270° < \theta < 360°$

$\alpha = 360 - \theta°$

$\sin \alpha = \dfrac{-y}{r}$; ∴ $\sin \theta = \dfrac{-y}{r}$

$\cos \alpha = \dfrac{x}{r}$; ∴ $\cos \theta = \dfrac{x}{r}$

$\tan \alpha = \dfrac{-y}{x}$; ∴ $\tan \theta = \dfrac{-y}{x}$

***Only* Cos is positive in Quadrant IV.**

In Quadrant IV: To find α, subtract θ from 360°.

A student friendly method of remembering which trig functions are positive in each of the quadrants is the saying, "All Students Take Calculus" where the first letter of each word represents the positive trig value in the quadrants I through IV. Since the two perpendicular axes form a + sign, that helps students remember the functions are + in this sketch.

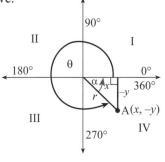

Sidebar: **Trigonometric Functions**

2.2

Examples

❶ If θ = 120°, what is the exact value of sin θ?

Solution: 120° is in quadrant II. Find the reference angle, α, by subtracting: 180° – 120° = 60°. The reference angle is 60°.

Since θ is in Quadrant II, its sine is positive $\sin 120° = \sin 60° = \dfrac{\sqrt{3}}{2}$.

❷ Find the tangent of an angle whose terminal side goes through the point (5, –6).

$$\tan \alpha = \frac{-6}{5}$$

$$\therefore \tan \theta = \frac{-6}{5}$$

Solution: Make a sketch of this one. θ is in Quadrant IV if the terminal side goes through (5, –6). Its reference angle is α.

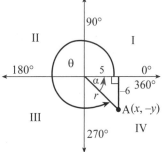

❸ Using the information from example 2, find the exact values of sin θ and cos θ.

Solution: The reference angle is still α, but we need the hypotenuse of the triangle. Find it with the Pythagorean Theorem and leave it in radical form since the question requires the exact value. Remember the hypotenuse, r, is always positive. Use r to find the sine and cosine of θ.

$$r^2 = (5)^2 + (-6)^2 \qquad \sin \alpha = \frac{-6}{\sqrt{61}} = \frac{-6\sqrt{61}}{61}$$

$$r = \sqrt{61}$$

$$\therefore \sin \theta = \frac{-6\sqrt{61}}{61}$$

$$\cos \alpha = \frac{5}{\sqrt{61}} = \frac{5\sqrt{61}}{61}$$

$$\cos \theta = \frac{5\sqrt{61}}{61}$$

❹ Using the value for cos θ found in example 3, find the measure of θ to the nearest tenth of a degree.

Solution: Use the inverse cosine function on the calculator to find a, the reference angle. Then subtract from 360° to find the value of θ in quadrant IV.

$\cos^{-1}\left(\dfrac{5\sqrt{61}}{61}\right) = 50.19$ ***This is the reference angle.***

$360° - 50.19° = 309.81°$, ***θ is in quadrant IV.***

$θ ≈ 309.8°$

❺ Convert the answer for example 4 to the nearest tenth of a radian.

Solution: Multiply the degrees by $\dfrac{\pi \text{ radians}}{180°}$.

$$309.8° \cdot \dfrac{\pi}{180°} = 5.4 \text{ radians}$$

Note: This problem directed us to use the answer from example 4, so we used the rounded value. Make sure and use the entire original value given in the calculator, or at least the value to several decimals places, in most problems.

SUMMARY

Memorize the special angles and quadrantal angles. (See page 88.) That knowledge, combined with the information on the diagram shown here will provide almost all the information you need to work with angles in various quadrants. The diagram summarizes finding reference angles α, the positive or negative values of the trig functions in the four quadrants, and includes the unit circle.

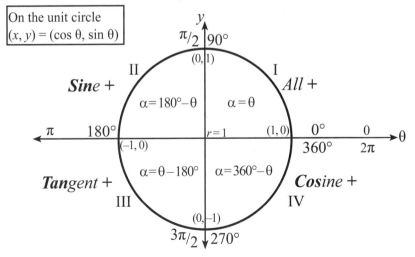

On the unit circle
$(x, y) = (\cos θ, \sin θ)$

Trigonometric Functions

UNIT CIRCLE AND TRIG GRAPHS

Unit Circle: A circle with a radius, r, equal to one that is used to work with rotational angles and trig functions. As the terminal side of the angle rotates counterclockwise, the (x, y) coordinates of the point of its intersection with the unit circle, (radius = 1), are the cosine and sine values of the central angle, θ, respectively. This relationship is true in all four quadrants and is also true for the quadrantal angles.

Trig Ratios In General

$$\cos \theta = \frac{x}{r} \qquad \sin \theta = \frac{y}{r} \qquad \tan \theta = \frac{y}{x}$$

On The Unit Circle

$$\cos \theta = \frac{x}{1} \qquad \sin \theta = \frac{y}{1} \qquad \tan \theta = \frac{y}{x}$$

$$\cos \theta = x \qquad \sin \theta = y \qquad \tan \theta = \frac{\sin \theta}{\cos \theta}$$

The coordinates of point A, the intersection of the terminal side of θ with the unit circle, are (x, y) or $(\cos \theta, \sin \theta)$.

The tangent of θ is found by dividing $\sin \theta$ by $\cos \theta$, or dividing the y coordinate of point A by the x coordinate of point A.

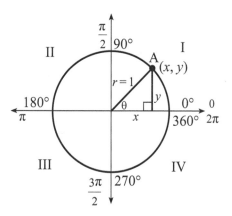

QUADRANTAL ANGLES

In this diagram, the coordinates of the points where the terminal side of the quadrantal angles intersect the unit circle are labeled. Remember, the *x* coordinate is the cosine value of the angle and the *y* coordinate is the sine value of the central angle.

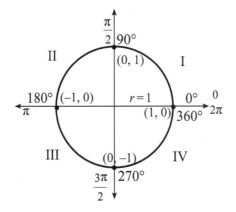

Central Angleθ in °and radians	$\cos\theta$ (x)	$\sin\theta$ (y)	$\tan\theta$ (y/x)
$0°, 0$	1	0	$\frac{0}{1} = 0$
$90°, \frac{\pi}{2}$	0	1	$\frac{1}{0}$ *undefined*
$180°, \pi$	-1	0	$\frac{0}{-1} = 0$
$270°, \frac{3\pi}{2}$	0	-1	$\frac{-1}{0}$ *undefined*
$360°, 2\pi$	1	0	$\frac{0}{1} = 0$

$0°$ or 0 radians and $360°$ or 2π share the same point of intersection with the unit circle. Their trig function values are the same.

Special Angles: On the unit circle, as the terminal side of θ, (in standard position), is rotated counterclockwise, the (x, y) values of each point of intersection with the unit circle are the cosine and sine values of the angle formed. Commonly, the angles that are multiples of 30°, 45°, 60° and the quadrantal angles are used to sketch the graphs of the trig functions.

θ in Radians	cos θ	sin θ	tan θ (sin θ/cos θ)
0 *	1	0	0
$\frac{\pi}{6}$	.8660	.5	.5774
$\frac{\pi}{4}$	.7071	.7071	1
$\frac{\pi}{3}$	.5	.8660	1.7321
$\frac{\pi}{2}$ *	1	undefined	0
$\frac{2\pi}{3}$	.8660	−1.7321	−.5
$\frac{3\pi}{4}$	−.7071	.7071	−1
$\frac{5\pi}{6}$	−.8660	.5	−.5774
π *	−1	0	0
$\frac{7\pi}{6}$	−.8660	−.5	.5774
$\frac{5\pi}{4}$	−.7071	−.7071	1
$\frac{4\pi}{3}$	−.5	−.8660	1.7321
$\frac{3\pi}{2}$ *	0	−1	undefined
$\frac{5\pi}{3}$	.5	−.8660	−1.7321
$\frac{7\pi}{4}$	.7071	−.7071	−1
$\frac{11\pi}{6}$	.8660	−.5	−.5774
2 π *	1	0	0
* Indicates a quadrantal angle			

When graphing, the x value of the point graphed represents the angle itself. It is usually in radian form but can be in degrees. Label accordingly. The angle can be represented by θ, x, or any upper case letter. The range, or y values, are the values of the trig functions. The points to be graphed will have the form $(\theta, \sin \theta)$, $(x, \sin x)$ or $(A, \sin A)$. The domain is $-\infty \le \theta \le \infty$ unless restricted.

Function	Domain	Range
Cosine	$-\infty \le \theta \le \infty$	$-1 \le y \le 1$
Sine	$-\infty \le \theta \le \infty$	$-1 \le y \le 1$
Tangent	$-\infty \le \theta \le \infty$ except odd multiples of $\frac{\pi}{2}$	$-\infty \le y \le \infty$

Periodic Functions: The values of the trig functions repeat in a pattern. These are also called cyclical functions.

Sine and Cosine Graphs: Both basic graphs have one complete cycle (one complete curve) within the domain of $0 \le \theta \le 2\pi$. The horizontal x-axis is the principal axis and is labeled with values of θ in radian form. The vertical axis is the y-axis and represents the value of the function at the appropriate value of θ. The y-axis is labeled with integers and each function has a maximum value of 1 and a minimum value of -1. They each have an amplitude of 1, a frequency of 1, and a period of 2π. (See page 91.)

Key Points: Each graph has five key values of θ for which the function is equal to a maximum value (1), minimum value (-1), or zero. The key values of θ for both functions are $0, \frac{\pi}{2}, \pi, \frac{3\pi}{2}, 2\pi$. The key points are located at these five values of θ and additional points in between them are plotted by estimating the y-values on the graph. Use the special angles for the additional points. A trig graph is labeled as shown. (Match the label on the horizontal axis to the variable representing the angle.)

Trigonometric Functions

Transformations performed on the basic graphs may change the location of the maximum, minimum, and zero values when the equations are graphed, but there will still be five key points in one cycle of each of the graphs. To determine the appropriate locations, we will find the five key points first.

- Graph of $f(x) = a \cos b(\theta + c) + d$ where $a = 1, b = 1, c = 0, d = 0$; $0 \le \theta \le 2\pi$

Key Points

(0, 1)

$\left(\dfrac{\pi}{2}, 0\right)$

$(\pi, -1)$

$\left(\dfrac{3\pi}{2}, 0\right)$

$(2\pi, 1)$

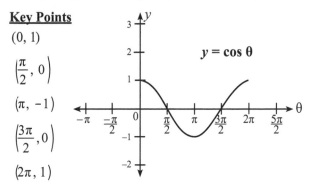

- Graph of $f(x) = a \sin b(\theta + c) + d$ where $a = 1, b = 1, c = 0, d = 0$; $0 \le \theta \le 2\pi$

Key Points

(0, 0)

$\left(\dfrac{\pi}{2}, 1\right)$

$(\pi, 0)$

$\left(\dfrac{3\pi}{2}, -1\right)$

$(2\pi, 0)$

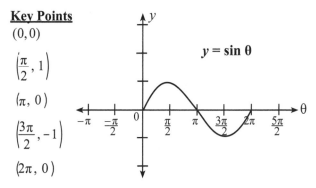

Due to the cyclical nature of trig graphs, both the cosine and sine graphs repeat their curves in both the negative and positive directions. The cosine graph is a translation (or shift) of the sine graph $\pi/2$ units to the left.

Expanded graph of $y = \cos \theta$

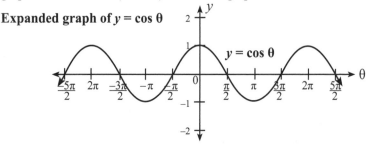

Expanded graph of $y = \sin \theta$

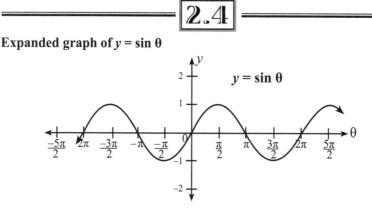

AMPLITUDE, FREQUENCY, AND PERIOD

The equations $y = a \sin b (x + c) + d$ or $y = a \cos b (x + c) + d$ provide additional information about trig graphs.

a represents the amplitude, b is the frequency, c horizontal shift, and d is the vertical shift.

In the basic graphs, a and b are both 1, and c and d are both zero. a and b involve the shape of the graph, the width and vertical measure of the curve. c and d involve placement of the curve on the axes.

Amplitude: One-half the vertical distance from the minimum point to the maximum point on the graph. $|a|$ is the value of the amplitude.

> **Example** $y = 3 \cos x$. The amplitude is 3. The maximum point of the graph is at $y = 3$, and the minimum point is at $y = -3$.

Frequency: The number of times one complete cycle or pattern of the function occurs within 2π. b is the frequency.

> **Example** $y = \cos 2x$. The frequency is 2. There are two complete cosine curves within 2π radians.

Period: The interval in degrees or radians that contains one complete cycle of the function. The period is found by dividing 2π by b.

> **Example** $y = \cos 2x$. b is 2 so the period is $2\pi/2$ or π. This means that one complete cycle of the curve occurs in π radians.

Trigonometric Functions

Sketching the Graphs

Examples

❶ $y = 3 \cos \theta \; ; \; 0 \le \theta \le 2\pi$

Amplitude: $a = 3$ Highest point is 3, lowest is –3.

Frequency: $b = 1$ One complete cycle occurs once in 2π.

Period: $\dfrac{2\pi}{1} = 2\pi$ Only one complete cycle appears in the interval 2π.

Key Points
(0, 3)

$(\dfrac{\pi}{2}, 0)$

$(\pi, -3)$

$(\dfrac{3\pi}{2}, 0)$

$(2\pi, 3)$

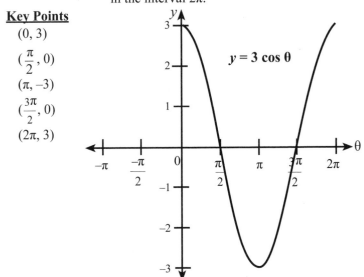

❷ $y = \sin (2\theta) \; ; \; -\pi \le \theta \le 2\pi$ (Note that the domain is $-\pi$ to 2π)

Amplitude: $a = 1$: Maximum point is 1, minimum is –1.

Frequency: $b = 2$: Two complete cycles will occur within 2π.

Period: $\dfrac{2\pi}{2} = \pi$: π is now the interval that contains a complete cycle.

Key Points
(0, 0)

$(\dfrac{\pi}{4}, 1)$

$(\dfrac{\pi}{2}, 0)$

$(\dfrac{3\pi}{4}, -1)$

$(\pi, 0)$

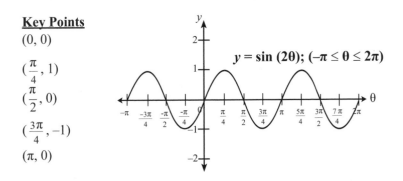

❸ $y = 3 \cos(\theta/2); \ 0 \le \theta \le 4\pi$
(Domain ends at 4π, label graph accordingly.)

Amplitude = 3. This will change the y values of the key points.
Maximum = 3, minimum = –3

Frequency = ½ One-half of the cycle is within 2π. This will change the x values of the key points.

Period = $2\pi/(1/2) = 4\pi$. One complete cycle will occur in 4π.

Key Points: The graph requires 4π to show one cycle instead of 2π. Double the independent variable, θ, of the basic key points. The maximum and minimum y values are now 3 and –3. Multiply the basic y values of the key points by 3.

<u>Key Points</u>

$(0, 3)$

$(\pi, 0)$

$(2\pi, -3)$

$(3\pi, 0)$

$(4\pi, 3)$

$$y = 3 \cos\left(\frac{\theta}{2}\right), \ (0 \le \theta \le 4\pi)$$

HORIZONTAL (PHASE) AND VERTICAL SHIFTS OF TRIG GRAPHS

$$y = a \sin b \, (x + c) + d$$
$$y = a \cos b \, (x + c) + d$$

In the equations shown in the previous section we used a and b to sketch the graph. (c and d were both zero.) c tells us where the graph is located horizontally on the principal axis or midline, and d tells use where the principal axis is located. (Remember that when $d = 0$, the principal axis is the x-axis.)

Phase Shift: A horizontal translation of the basic graph which is indicated by c.

- When c is *positive*, the graph (and its key points) move $|c|$ units to the LEFT.

- When c is *negative*, the graph moves $|c|$ units to the RIGHT.

- The phase shift is described as being equal to $(-c)$.

Key Points: To locate the x value of the key points, subtract c from the x (independent variable) value of the basic key points.

Example $y = \cos(\theta - \pi/4)$. Graph in the interval $-\pi \le \theta \le 2\pi$. Start with the key points of the basic graph between 0 and 2π. Since $b = 1$, subtract $-\pi/4$ from each x value. (Add $\pi/4$). Plot. Then extend repeating pattern to include the interval needed.

Key points	$\left(\text{Basic } \theta - \left(\dfrac{-\pi}{4}\right), y \right) \Rightarrow \left(\text{\textit{Shifted} } \theta, y \right)$
$(0, 1)$	$\left(0 + \dfrac{\pi}{4}, 1 \right) \Rightarrow \left(\dfrac{\pi}{4}, 1 \right)$
$\left(\dfrac{\pi}{2}, 0 \right)$	$\left(\dfrac{\pi}{2} + \dfrac{\pi}{4}, 0 \right) \Rightarrow \left(\dfrac{3\pi}{4}, 0 \right)$
$(\pi, -1)$	$\left(\pi + \dfrac{\pi}{4}, -1 \right) \Rightarrow \left(\dfrac{5\pi}{4}, -1 \right)$
$\left(\dfrac{3\pi}{2}, 0 \right)$	$\left(\dfrac{3\pi}{2} + \dfrac{\pi}{4}, 0 \right) \Rightarrow \left(\dfrac{7\pi}{4}, 0 \right)$
$(2\pi, 1)$	$\left(2\pi + \dfrac{\pi}{4}, 1 \right) \Rightarrow \left(\dfrac{9\pi}{4}, 1 \right)$

Plot the shifted points. Extend the pattern to the left to include $-\pi$.

$y = \cos\left(\theta - \dfrac{\pi}{4} \right), \ (-\pi \le \theta \le 2\pi)$

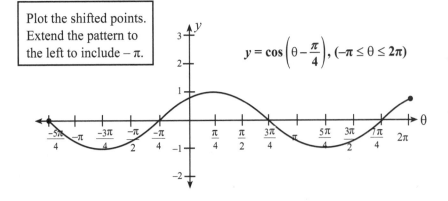

Vertical Shift: *d* in the equation indicates a vertical translation of the trig graph. The principal axis for the basic graph is the *x*-axis, or *y* = 0. The principal axis for the vertically shifted graph will have its principal axis at *y* = *d*. *d* must be added to the *y* values of the key points.

Example $y = \sin(x) - 3$,

Key points	(Basic **x**, **y** + **d**) ⇒ (**x**, **shifted y**)
$(0, 0)$	$\left(0, 0-3\right) \Rightarrow (0, -3)$
$(\frac{\pi}{2}, 1)$	$\left(\frac{\pi}{2}, 1-3\right) \Rightarrow \left(\frac{\pi}{2}, -2\right)$
$(\pi, 0)$	$\left(\pi, 0-3\right) \Rightarrow \left(\pi, -3\right)$
$(\frac{3\pi}{2}, -1)$	$\left(\frac{3\pi}{2}, -1-3\right) \Rightarrow \left(\frac{3\pi}{2}, -4\right)$
$(2\pi, 0)$	$\left(2\pi, 0-3\right) \Rightarrow \left(2\pi, -3\right)$

Plot the shifted points, then extend the repeating pattern to -2π.

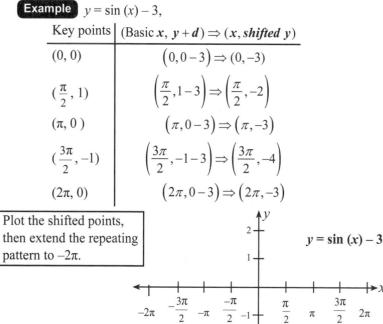

$y = \sin(x) - 3$

midline

Examples

❶ The temperatures, in Fahrenheit degrees, in a New England town, starting January 1, can be modeled by the function $f(x) = -30\cos\frac{\pi}{6}x + 40$. Graph the function on the grid below in the interval $0 \le x \le 24$ and discuss the characteristics of the graphed function. Include the midline, amplitude, maximum and minimum points, the period of the function, and any additional characteristics of interest. Write a paragraph describing these characteristics in terms of the temperature conditions in the town.

Solution: Input the function in the calculator and graph it. Use the calculations for the minimum and maximum points. Determine the midline by locating the line half way between the maximum and minimum points. Then discuss the months and temperatures as they relate to the town.

Temperature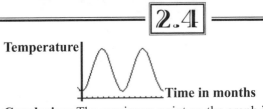

Time in months

Conclusion: The maximum point on the graph is (6, 70) and the minimum point is (12, 10).

The amplitude is $|a| = |-30| = 30$. The period is $\dfrac{2\pi}{\frac{\pi}{6}} = 12$

The midline is $y = \dfrac{Max\ y + Min\ y}{2} = \dfrac{70 + 10}{2} = 40$.

The graph is a cosine function that has been reflected over its midline, $y = 40$.

Discussion: The function begins at its minimum value of 10 which means that when the record keeping began, it was January 1 and the temperature was 10° F. The temperature increased as time continued through the months and in the 6th month, June, the maximum temperature of 70° F was reached. After that, the temperature fell until it reached its lowest point at the end of 12 months, December 31. The average temperature for the year was 40°F and the temperatures ranged from 10°F to 70°F. One complete cycle to the temperatures occurred in 12 months.

❷ In an amusement park a kiddie ride with swings operates by rotating the swings in a circle and moving them higher and lower in relation to the ground. The length of the arm from the center of the ride to the swing is 20 feet. The height of a swing, in feet, can be modeled by the function $f(t) = 3 \sin \frac{\pi}{10} t + 7$ where t represents the seconds after the ride started. Graph the function. What is the maximum height and the minimum height of the swing measured from the ground. At what height does the swing begin? How long does it take for the swing to its maximum height back to its maximum height again? About how many cycles are completed in one minute?

Solution: Graph using an appropriate window in the calculator. Sketch the graph. Use the maximum and minimum values to determine the heights and the time between the maximum and minimum.

Height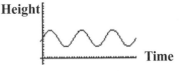

Time

Conclusion:
The maximum height of the swing is about 10 feet and it is reached 5 seconds after the ride starts. The minimum height is about 4 feet and it is reached after 15 seconds on the ride. The swing begins at a height of 7 feet. The time for one complete cycle, from maximum height to maximum height again is 20 seconds, so three cycles will occur within a minute.

Algebra 2 Made Easy – Common Core Edition

Tangent Graph:

$y = a \tan bx$ Domain $-\infty \leq \theta \leq \infty$, except odd multiples of $\frac{\pi}{2}$.

Remember that: $\tan x = \frac{\sin x}{\cos x}$. The domain of the graph is determined by this.

- When $\cos x = 0$, the tangent function is undefined. Tangent is undefined at odd multiples of $\frac{\pi}{2}$. At those values of x, a vertical asymptote indicates the line that the y values approach positive or negative infinity. The *domain* of the tangent function is $\{x: \text{R}, x \neq n \cdot \frac{\pi}{2}$ where n is an odd integer$\}$
- When $\sin x = 0$, the tangent function $= 0$. This happens at ... , $-\pi$, 0, π, 2π, ...

Amplitude: Tangent Graph has no amplitude. Its _range_ is $-\infty \leq y \leq \infty$.

Period: π. This is a characteristic of the tangent graph. It is not determined by dividing $\frac{2\pi}{b}$ as it is in the sine and cosine graphs. Two tangent cycles appear within an interval of 2π. Use $\frac{\pi}{b}$ to find the period when b is not 1.

Key Points: The tangent graph has 3 points that are easy to graph and as the graph approaches the vertical asymptotes, it approaches positive or negative infinity.

Examples

❶ The key points of the basic graph are

as x approaches $-\frac{\pi}{2}$, $y \to -\infty$

$(-\frac{\pi}{4}, -1)$

$(0, 0)$

$(\frac{\pi}{4}, 1)$

and as x approaches $\frac{\pi}{2}$, $y \to \infty$

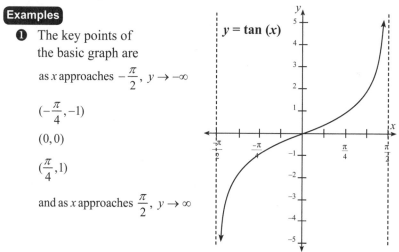

Note: An asymptote is a line that a graph approaches more and more closely.

The vertical asymptotes are the dashed lines. They repeat throughout the graph at odd multiples of $\frac{\pi}{2}$.

Trigonometric Functions

❷ $y = \tan 2x$, $(-\pi \leq x \leq \pi)$

Asymptotes are at $\dfrac{-\pi}{4}$ and $\dfrac{\pi}{4}$.

Period on this graph $\dfrac{\pi}{2}$. One complete cycle occurs in an interval of $\dfrac{\pi}{2}$.

Key Points

As x approaches

$\dfrac{-\pi}{4}$, $y \to -\infty$

$(\dfrac{-\pi}{8}, -1)$

$(0, 0)$

$(\dfrac{\pi}{8}, 1)$

As x approaches

$\dfrac{\pi}{4}$, $y \to \infty$

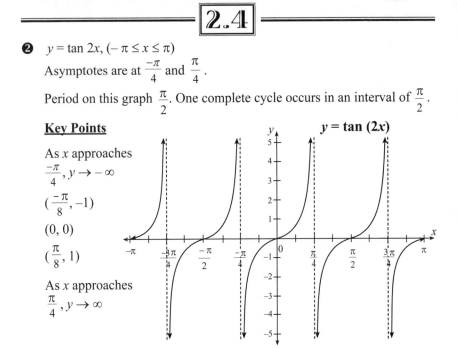

$y = \tan(2x)$

❸ $y = 3 \tan(x)$, $(-2\pi \leq x \leq 2\pi)$

Asymptotes are at $x = \dfrac{-\pi}{2}$ and $\dfrac{\pi}{2}$.

Period: π

Key Points

As x approaches

$\dfrac{-\pi}{2}$, $y = \to -\infty$

$(\dfrac{-\pi}{4}, -3)$

$(0, 0)$

$(\dfrac{\pi}{4}, 3)$

As x approaches

$\dfrac{\pi}{2}$, $y = \to \infty$

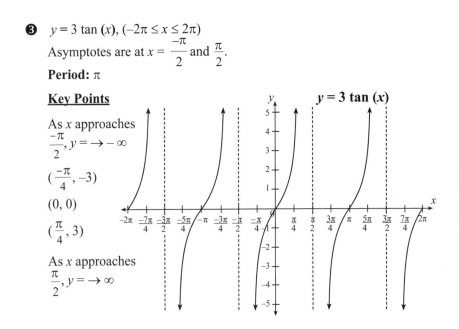

$y = 3 \tan(x)$

GRAPHING INVERSE TRIG FUNCTIONS

Remember, a function must be one-to-one for it to have an inverse function.

To find the inverse of a function, the (x, y) coordinates of the points in the function are reversed and then they become the points on the inverse. If a point on $f(x)$ is $(2, 7)$, the corresponding point on $f^{-1}(x)$, the inverse, is $(7, 2)$. The graph of $f^{-1}(x)$ is the reflection of the graph of $f(x)$ over the line $y = x$.

Domain and Range of trig functions and their inverses. The domain of the function is the range of its inverse. The range of the function is the domain of its inverse. In the function, the domain shows the possible values of the angle. The range is the y values of the trig function. In the inverse, the domain is the possible values of the trig function, and the angles values are the range.

The domain of the sine, cosine, and tangent functions must be restricted.

Examples

❶ *Function*: $sin\ x$ *Inverse* : *arc sin x or* $sin^{-1}x$

  Domain $x : -\dfrac{\pi}{2} \le x \le \dfrac{\pi}{2}$ $x : -1 \le x \le 1$

  Range $y : -1 \le y \le 1$ $y : -\dfrac{\pi}{2} \le y \le \dfrac{\pi}{2}$

  $f(x) = \sin(x)$ $g(x) = \sin^{-1}(x)$

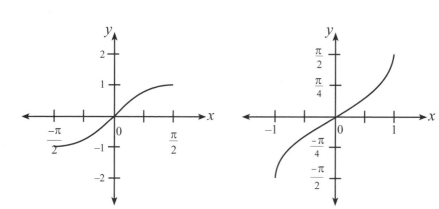

Trigonometric Functions

❷ Function: *cos x*

Domain $x : 0 \le x \le \pi$

Range $y : -1 \le y \le 1$

Inverse: $\cos^{-1} x$ *or arc cos x*

$x : -1 \le x \le 1$

$y : 0 \le y \le \pi$

$f(x) = \cos(x)$

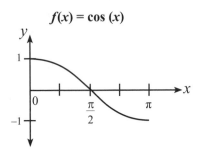

$g(x) = \cos^{-1}(x)$

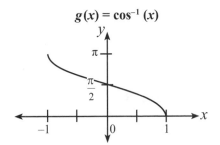

❸ Function: *tan x*

Domain $x : -\dfrac{\pi}{2} < x < \dfrac{\pi}{2}$

Range $y : -\infty \le y \le \infty$
(all real numbers)

Inverse: $\tan^{-1} x$ *or arc tan x*

$x : -\infty \le x \le \infty$ *All real numbers.*

$y : -\dfrac{\pi}{2} < y < \dfrac{\pi}{2}$

$f(x) = \tan(x)$

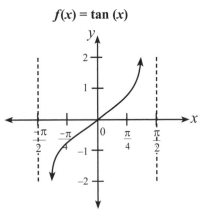

$g(x) = \tan^{-1}(x)$

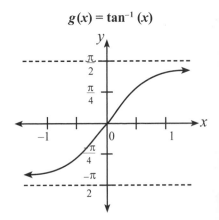

2.6

Do these graphs in the calculator and then transfer sketches to graph paper.

Examples

❶ Cosecant x: $\dfrac{1}{\sin x} = \csc x$

Domain: Reals except for multiples of π, where $\sin x = 0$

Range: $(y \le -1)$ or $(y \ge 1)$

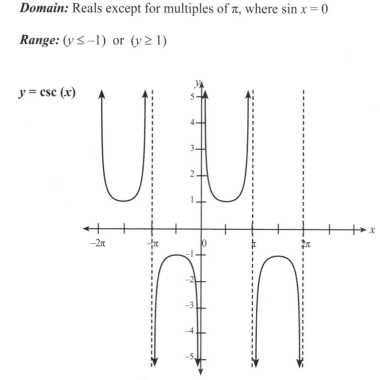

$y = \csc(x)$

Trigonometric Functions

❷ Secant x: $\dfrac{1}{\cos x} = \sec x$

Domain: R, except odd multiples of $\dfrac{\pi}{2}$, where $\cos x = 0$

Range: $(y \le -1)$ or $(y \ge 1)$

$y = \sec (x)$

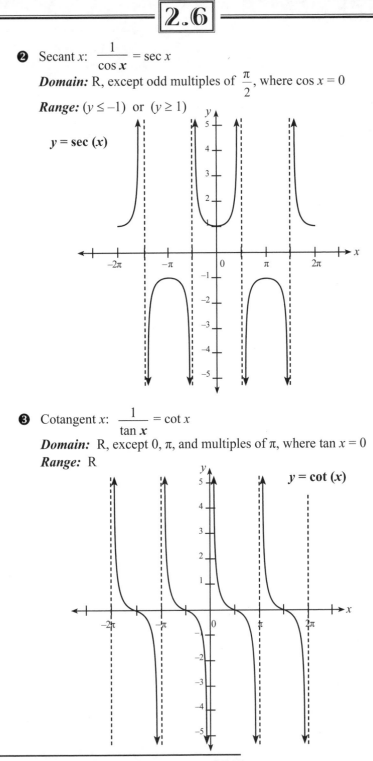

❸ Cotangent x: $\dfrac{1}{\tan x} = \cot x$

Domain: R, except 0, π, and multiples of π, where $\tan x = 0$

Range: R

$y = \cot (x)$

TRIG IDENTITIES AND EQUATIONS

A trigonometric identity, ID, is an equation that is true no matter what the value of the independent variable. Trig ID's are substituted into more complex trig expressions to simplify them and to solve equations.

Reciprocal Identities

$$\sec \theta = \frac{1}{\cos \theta} \qquad\qquad \cos \theta = \frac{1}{\sec \theta}$$

$$\csc \theta = \frac{1}{\sin \theta} \qquad\qquad \sin \theta = \frac{1}{\csc \theta}$$

$$\cot \theta = \frac{1}{\tan \theta} \qquad\qquad \tan \theta = \frac{1}{\cot \theta}$$

$$\tan \theta = \frac{\sin \theta}{\cos \theta} \qquad\qquad \cot \theta = \frac{\cos \theta}{\sin \theta}$$

Examples Rewrite these expressions as a simplified fraction in terms of $\sin \theta$ or $\cos \theta$ or both.

❶ $\csc \theta - \cot \theta \Rightarrow \dfrac{1}{\sin \theta} - \dfrac{\cos \theta}{\sin \theta} = \dfrac{1 - \cos \theta}{\sin \theta}$

❷ $\dfrac{\cot \theta}{\csc \theta} \Rightarrow \dfrac{\dfrac{\cos \theta}{\sin \theta}}{\dfrac{1}{\sin \theta}} = \dfrac{\cos \theta}{\sin \theta} \cdot \dfrac{\sin \theta}{1} = \cos \theta$

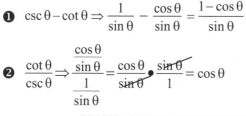

PYTHAGOREAN IDENTITIES

The Pythagorean Identities are derived from the trig relationships in the unit circle. The coordinates of the point of intersection of the terminal side of angle θ, and the unit circle are (x, y) and are equivalent to $(\cos \theta, \sin \theta)$. When a right triangle is drawn with the terminal side of the angle as the hypotenuse, one leg of the triangle formed is equal to x, the other leg equals y. The radius, r, is 1 in the unit circle.

Substituting x, y and the value of r into the Pythagorean Theorem: $\quad y^2 + x^2 = 1$

Since $x = \cos \theta$ and $y = \sin \theta$, the Pythagorean Identity for trig is: $\boxed{\sin^2 \theta + \cos^2 \theta = 1}$

This identity can be rearranged algebraically into two more forms using sine and cosine: $\boxed{\sin^2 \theta = 1 - \cos^2 \theta}$ and $\boxed{\cos^2 \theta = 1 - \sin^2 \theta}$

Trigonometric Functions

If the main Pythagorean Identity is divided by $\sin^2 \theta$ or by $\cos^2 \theta$, several additional forms can be created algebraically. They involve the reciprocal identities.

$$\sin^2 \theta + \cos^2 \theta = 1 \qquad \text{and} \qquad \sin^2 \theta + \cos^2 \theta = 1$$

$$\frac{\sin^2 \theta}{\sin^2 \theta} + \frac{\cos^2 \theta}{\sin^2 \theta} = \frac{1}{\sin^2 \theta} \qquad\qquad \frac{\sin^2 \theta}{\cos^2 \theta} + \frac{\cos^2 \theta}{\cos^2 \theta} = \frac{1}{\cos^2 \theta}$$

$$\Downarrow \qquad\qquad\qquad\qquad\qquad \Downarrow$$

$$\boxed{1 + \cot^2 \theta = \csc^2 \theta} \qquad\qquad \boxed{\tan^2 \theta + 1 = \sec^2 \theta}$$

or $\qquad\qquad\qquad\qquad\qquad$ *or*

$$\boxed{\cot^2 \theta = \csc^2 \theta - 1} \qquad\qquad \boxed{\tan^2 \theta = \sec^2 \theta - 1}$$

or $\qquad\qquad\qquad\qquad\qquad$ *or*

$$\boxed{1 = \csc^2 \theta - \cot^2 \theta} \qquad\qquad \boxed{1 = \sec^2 \theta - \tan^2 \theta}$$

Summary: The Pythagorean Identity, $\boxed{\sin^2 \theta + \cos^2 \theta = 1}$, can be used in nine different forms. It is informally referred to as the "Magic Rule".

Examples Simplify each expression to a single trig function.

❶ $\quad \sin x(1 + \cot^2 x) \qquad$ [***Hint:*** Substitute $\csc^2 x$ for $(1 + \cot^2 x)$]

$$\sin x(\csc^2 x) = (\sin x)\left(\frac{1}{\sin^2 x}\right) = \frac{1}{\sin x} \quad \text{or} \ \csc x$$

❷ $\quad 3 - 3 \sin^2 x \qquad$ $\left[\begin{array}{l}\textbf{\textit{Hint:}} \text{ Factor the GCF of 3, then substitute} \\ \qquad \cos^2 x \text{ for } (1 - \sin^2 x)\end{array}\right]$

$$3(1 - \sin^2 x) = 3\cos^2 x$$

❸ $\quad \dfrac{3}{\cos^2 x} - 3 \qquad$ $\left[\begin{array}{l}\textbf{\textit{Hint :}} \text{ GCF is 3, then use reciprocal ID for } \dfrac{1}{\cos^2 x} \\ \quad \text{Substitute } \tan^2 x \text{ for } (\sec^2 x - 1).\end{array}\right]$

$$3\left(\frac{1}{\cos^2 x} - 1\right) = 3(\sec^2 x - 1) = 3(\tan^2 x) = 3\tan^2 x$$

Unit 3

FUNCTIONS

- Understand the concept of a function and use function notation.

- Interpret functions that arise in applications in terms of the context.

- Analyze functions using different representations.

- Write a function in different forms to explain different properties of the function.

- Build a function that models a relationship between two quantities.

- Build a new function from existing functions.

- Construct and compare linear, quadratic, and exponential models and solve problems.

- Interpret expressions for functions in terms of the situation they model.

FUNCTIONS

Function and relation are terms that can be used to name equations or rules in higher math. Both refer to a set of ordered pairs that are associated to each other in a way that is expressed by the relation or the function. Functions and relations both have domains and ranges.

Relation: A set of ordered pairs that have a connection to each other that is defined by the relation or rule. The ordered pairs may be numeric or not.

> **Examples** ❶ (2, 4), (6, 8), (6, 4)
> ❷ (Albany, NY), (Boston, MA), (Sacramento, CA)

Function: A function is a relation that consists of a set of ordered pairs in which each value of x is connected to a unique value of y based on the rule of the function. For each x value, there is one and only one corresponding value of y.

> **Example** (2, 3), (3, 4), (4, 5), (5, 4), (6, 4)

Note: No x's are repeated, but y can be repeated.

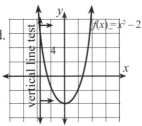

- **Vertical Line Test:** When a function is graphed, a vertical line passed across the graph will intersect the function graph in only one point at a time.

Notation for function is $f(x)$. Any letter can name the function, and the independent variable is noted in the ( ). For instance, $h(x)$, or $g(x)$. This notation replaces y in the usual equation format. Instead of $y = x + 3$, it is $f(x) = x + 3$.

Domain: The largest set of elements available for the independent variable, the first member of the ordered pair (x). The domain is the set of Real Numbers unless otherwise noted. Restrictions on the domain can be found when fractions or radicals are involved.

- **Fraction:** Denominator cannot be zero. $f(x) = \dfrac{x-4}{x+3}, x \neq -3$

- **Radical:** Randicand cannot be negative. $f(x) = \sqrt{4x - 10}; \ x \geq \dfrac{5}{2}$

Range: The set of elements for the dependent variable, the second member of the ordered pair (y). The range can be found for the entire relation or function, or it can be found for specific domains. If a value of x is given as the domain, substitute to find the corresponding value of y (range).

Algebra 2 Made Easy – Common Core Edition

Working with the Domain and Range:

Examples

❶ Find the domain and range of the equation $f(x) = x + 4$. The domain (values available for x) is R, (real numbers). The range in this case (numbers available for y) is also R. (There are no restrictions on this function.)

❷ $y = x^2$ Domain is the Reals, range is $y \geq 0$.
(any real number squared is positive)

❸ If the domain is $\{-3, 0, 4\}$ find the range of $y = x - 3$
Substitute -3: $y = (-3) - 3 = -6$
Substitute 0: $y = (0) - 3 = -3$
Substitute 4: $y = 4 - 3 = 1$
Note: In this relation, if the domain is $\{-3, 0, 4\}$ the range is $\{-6, -3, 1\}$.

One to One Function: A 1-1 must be a function. Secondly, when the ordered pairs are examined, no two of them have the same y value. No x's can be repeated, and no y's can be repeated.

In a one-to-one function, a vertical line test works, and also a horizontal line passed over the graph will intersect the graph in only one point at a time.

Examples

❶ $(2, 3), (3, 4), (4, 5)$ is $1 - 1$.
$(2, 3), (4, 3), (5, 6)$ is not $1 - 1$ because 3 is repeated as a y value.

❷ Set M is the domain, x, and it contains all the real numbers. Set C is the range, y, of the function and contains all real numbers. $f: M \rightarrow C$ Each value of x has a unique value for y, and each y has a unique value for x.

❸ In the figure to the right, the curve represents $f(x)$. It is $1 - 1$ because the vertical line test and the horizontal line test both work.

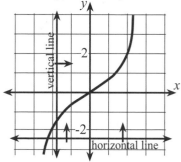

Building and Interpreting Functions

107

Finding the Domain: Remember that the domain is always the set of real numbers unless specified otherwise or restricted by the contents of the relation or function because of a fraction or radical or both! It is the largest set of numbers that will result in a real number solution.

To find restrictions, determine what values of x will not give a real number output. (Since the "default" set is all the real numbers, sometimes just the restrictions are written to indicate the domain.)

Examples

❶ $y = \dfrac{2x + 3}{2x - 1}$

In a fraction: determine which values of x make the denominator $= 0$.

$\therefore 2x - 1 \neq 0$

$x \neq \dfrac{1}{2}$

Domain: $\{R, x \neq \dfrac{1}{2}\}$

❷ $\sqrt{x - 3}$

$x - 3 \geq 0$

$x \geq 3$

In an even index radical: determine which values of x make the radicand non-negative.

Domain: $\{R, x \geq 3\}$

❸ $\dfrac{7 - x}{\sqrt{x - 5}}$

$x - 5 > 0$

$x > 5$

In a fraction with an even index radical in the denominator: determine which values of x make the denominator zero and which values make any radicand in the fraction positive.

Domain: $\{R, x > 5\}$

Practice: Find the domain and the range where possible. It is often helpful to use a graph to find the range. Determine if the equation is a function or not, and if it is a function, is it one-to-one?

Examples

❶ $y = \dfrac{3}{|2x - 7|}$

$2x - 7 \neq 0$

$2x \neq 7, \qquad x \neq \dfrac{7}{2}$

Domain: $\left\{ x : x \neq \dfrac{7}{2} \right\}$

Range: $\{ y : y > 0 \}$

The graph shows the domain and range. The horizontal line test does not work so it is not one-to-one.

The vertical line test works:

$y = \dfrac{3}{|2x - 7|}$ is a function.

❷ $y = \sqrt{3x - 2}$

$3x - 2 \geq 0$

$x \geq \dfrac{2}{3}$

Domain: $\left\{ x : x \geq \dfrac{2}{3} \right\}$

Range: $\{ y : y \geq 0 \}$

It passes the vertical line test and the horizontal line test. This is a one-to-one function.

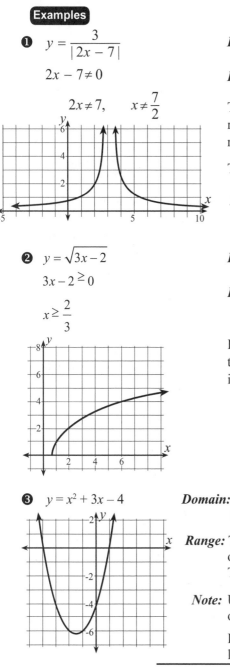

❸ $y = x^2 + 3x - 4$

Domain: No restrictions. The domain is the set of real numbers.

Range: The minimum point on the graph of this equation is $(-1.5, -6.25)$. The range is $\{ y : y \geq -6.25 \}$

Note: Use 2nd CALC 3:*minimum* on the calculator.

It is a function because the vertical line test works, not one-to-one.

4 $y = \dfrac{5}{\sqrt{x^2 - 4}}$

Domain: $\{x : (x < -2) \vee (x > 2)\}$

$x^2 - 4 > 0$

Range: $\{y : y > 0\}$

$(x < -2) \vee (x > 2)$

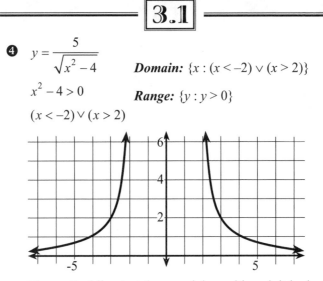

The vertical line test does work here although it looks like it might not. The graph is actually getting closer and closer to $x = -2$ and to $x = 2$ as y increases, so the vertical line will not intersect more than one point at a time.

It is a function.

Asymptote: When the graph approaches a line more and more closely, the line is called an asymptote. In the graph above, $x = 2$ and $x = -2$ are asymptotes.

Test Yourself: Test whether these are functions or not by viewing the graph. Find the domain. Are any of them 1-1?

5 $\dfrac{\sqrt{2 + x}}{\sqrt{x - 5}}$

Domain: _____

$1 - 1? :$ _____

6 $\dfrac{5}{x^2 - x - 6}$

Domain: _____

$1 - 1? :$ _____

7 $\dfrac{1 + \dfrac{1}{x}}{x - 5}$

Domain: _____

$1 - 1? :$ _____

(Answers to questions 5 –7 are given on next page)

Algebra 2 Made Easy – Common Core Edition

What is the domain of each?	Is it 1 – 1?
5) **Domain:** $x > 5$	5) 1-1 function
6) **Domain:** $x \neq 3, -2$	6) Function. x can't be 3, or -2. The vertical lines that appear on some calculators are the asymptotes of the graph and not part of the graph itself.
7) **Domain:** $x \neq 0, 5$	7) Function.

Examples

❶ The cost to operate a car includes gas price, insurance cost, finance charges, and maintenance costs. Charlie knows the cost to operate his car is approximated at $0.769 per mile. He sells his car when it has 97,000 miles on it. Using function notation write a function, C, to represent the cost of operating the car in terms of the miles driven, m. Determine the amount it has cost Charlie to drive his car when he sells it. What is a reasonable domain for this function? Explain your reason.

Solution: In general the function C equals the cost per mile times the number of miles driven. In function notation the cost is $f(m) = 0.769\ m$. When he sells the car it has 97,000 miles on it. Find $f(97,000)$ to determine how much it cost Charlie to operate the car.

$$f(97,000) = 0.769(97,000) \Rightarrow f(97,000) = 74,593$$

Conclusion: The function is $f(m) = 0.769\ m$. Charlie's cost to operate his car for 97,000 miles was $74,593. A reasonable domain is $x \geq 0$ since he cannot drive the car a negative amount of miles.

❷ Suzie is planning a trip to Australia in February. She will visit 5 cities. When Suzie starts to pack her suitcase, she doesn't know what kind of clothes to take. She looks online at international weather websites and finds out that the average temperatures for Australia are given in Celsius degrees, not Fahrenheit. Fortunately, from science class Suzie remembered that the formula $F = \frac{9}{5} C + 32$. Write the formula using function notation. Using the table below and the function, find the average high temperatures that are expected in the cities she will visit. What type of clothing should she pack?

Solution: The function to represent this formula is $F = \frac{9}{5} C + 32$.

City	Avg. High C° Temp.	$F(x) = \frac{9}{5}x + 32$	Avg. F° Temp
Perth	30	$F(30) = \frac{9}{5}(30) + 32$	86
Sydney	27	$F(27) = \frac{9}{5}(27) + 32$	80.6
Melbourne	26	$F(x) = \frac{9}{5}(26) + 32$	78.8
Brisbane	28	$F(x) = \frac{9}{5}(26) + 32$	82.4

Conclusion: The temperatures should be approximately 80°F at the time she visits. She needs to pack summer clothes.

Building and Interpreting Functions

GRAPHS OF FUNCTIONS

End behavior is defined as the trend of the value of $f(x)$ as x approaches ∞ or $-\infty$. End behavior can be determined by looking at the graph of the function. Although it isn't possible to graph all the way to the extremes of the x-axis, the trend of the values of $f(x)$ can be predicted. End behavior does not describe the middle parts of the graphed function.

Terminology used to describe end behavior is varied. The symbol $\rightarrow$ means "approaches." Word descriptions may include upward, rising, or approaching infinity ($\rightarrow \infty$); downward, falling, or approaching negative infinity ($\rightarrow -\infty$). In cases where the graph has an asymptote the end behavior can be described as approaching the asymptote (e.g. Right side $\rightarrow$ positive x-axis). The chart below describes the end behavior of functions in the form $f(x) = ax^n$.

In a *quadratic function* or another polynomial function with an even degree (highest exponent, n, is 2, 4, 6,...) the graph has two "arms" – one on the left and one on the right. The end behavior is either both arms rising or both arms falling. The positive or negative value of the coefficient, a, of the x^2 term determines whether upward or downward behavior occurs.
See Figures 1 and 2.

A third degree function, also called a **cubic function**, has "arms" that go in opposite directions. One is rising upward and the other is falling downward as x changes. The behavior depends on the positiove or negative value of a, the coefficient of the x^3 term. This end behavior also occurs for other functions where n, the exponent, is odd (3, 5, 7, ...). **See Figures 3 and 4**.

Value of a	Positive	Negative	Positive	Negative
Value of n	Even	Even	Odd	Odd
	Figure 1 $a > 0$	**Figure 2** $a < 0$	**Figure 3** $a > 0$	**Figure 4** $a < 0$
Graph				
End Behavior on *left* side of graph where $x \rightarrow -\infty$	$f(x) \rightarrow \infty$ Graph rises	$f(x) \rightarrow -\infty$ Graph falls	$f(x) \rightarrow -\infty$ Graph falls	$f(x) \rightarrow \infty$ Graph rises
End Behavior on *right* side of graph where $x \rightarrow \infty$	$f(x) \rightarrow \infty$ Graph rises	$f(x) \rightarrow -\infty$ Graph falls	$f(x) \rightarrow \infty$ Graph rises	$f(x) \rightarrow -\infty$ Graph falls

Algebra 2 Made Easy – Common Core Edition

The graph of an exponential function is in the form $f(x) = n^x$ where n is positive, some special characteristics are present. When $n > 1$ the graph is increasing and approaches the x-axis as x approaches negative infinity. If $0 < n < 1$ the graph is decreasing and approaches the x-axis as x approaches positive infinity. Both have vertical asymptotes that depend on the value of the base of the exponent. **See Figures 5 and 6**.

Special Note: Remember that when an exponent is negative, the base used is the reciprocal of the given base raised to the positive value of the exponent. The exponential function $f(x) = 5^{-x}$ is equivalent to $f(x) = \dfrac{1}{5^x}$ thus changing the value of n from $n > 1$ to $0 < n < 1$. That changes the graph from increasing to decreasing.

Likewise if $0 < n < 1$ is the base with a negative value of x, using the reciprocal with a positive exponent makes the base, $n > 1$, and the graph is then increasing. $f(x) = -x$ or $f(x) = .5^{-x}$ is equivalent to $f(x) = 2x$.

End Behavior of Graphs With Asymptotes: Exponential functions when graphed have a line which one arm of the graph approaches but does not cross. This is called an asymptote.

Value of n	$n > 1$	$0 < n < 1$
	Figure 5	**Figure 6**
Graph	$f(x) = 2^x$	$f(x) = .5^x$
End Behavior on *left* side of graph where $x \to -\infty$	Graph approaches the negative x-axis. $f(x) \to 0$	Graph is rising, $f(x) \to \infty$
End Behavior on *right* side of graph where $x \to \infty$	Graph is rising, $f(x) \to \infty$	Graph approaches the positive x-axis. $f(x) \to 0$

Examples

❶ Which of the following functions decreases as the independent variable values approach both positive and negative infinity?

1) $y = x^4 - 2x^2 + 2x + 5$ 3) $y = -x^4 + 3x^3 + x^2 + 5$
2) $y = x^3 - 3x^2 - x$ 4) $y = -2x^3 + 4x^2 + 6$

Solution: Use the graphing calculator and input each equation. The end behavior can be determined by examining the graph. (There is a rule about the degree of the variable – if you know it, use it! See page 112.)
Conclusion: The function $y = -x^4 + 3x^3 + x^2 + 5$ has end behaviors that decrease on both the positive and negative sides.

Building and Interpreting Functions

❷ Pete and Jack are doing a science experiment and part of the project is to graph the results of each of their experiments on the same graph and to describe the differences between them. Pete's experiment can be represented by the function $f(x) = (0.7)^x$. Jack's project is represented by $f(x) = (0.7)^x + 4$. Graph both functions on the grid below and discuss their differences. Include the intercepts, end behaviors, intersections, and explain why the graphs are in different positions.

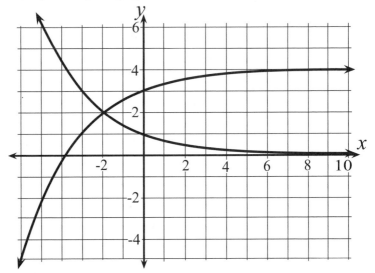

Solution: Graph both functions using the graphing calculator. Use the "calculate" function to determine the most accurate coordinates of the points of interest.

Discussion: Pete's function, $f(x)$, rises as the independent variable, x, approaches negative infinity and decreases as x approaches positive infinity. Jack's function, $g(x)$, has end behaviors that are opposite Pete's. This is caused by the negative sign in $g(x)$ which indicates a reflection of the graph of $f(x)$ over the x axis.

The y-intercept of $f(x)$ is at $y = 1$ and the y-intercept of $g(x)$ is at $y = 3$. After the reflection of $f(x)$, the graph is translated 4 units up to create $g(x)$. This causes the y-intercept to move from $y = -1$ to $y = 3$. There is no x-intercept of $f(x)$ and $y = 0$ is an asymptote of this graph. The x-intercept of $g(x)$ is at $x = -4$. An asymptote of $g(x)$ is $y = 4$.

The coordinates of the point of intersection are approximately $(-1.943, 2)$ as determined by the calculator although visually the graph appears to intersect at $(-2, 2)$.

ODD AND EVEN FUNCTIONS

The symmetry of a graph that represents a function can be described as odd, even, or neither. This can be determined by examining the graph or working with the function in its algebraic form.

Even Function: Algebraically a function is even if, for every number x in the domain, the number $-x$ is also in that domain. In function notation a function is even if $f(-x) = f(x)$. When viewed on a graph of a function, an even function is represented by a graph that is symmetric to the y-axis.

Odd Function: A function is odd if, and only if, the point $(+x, +y)$ is on the graph and the point $(-x, -y)$ is also on the graph. An odd function is defined as follows: A function is odd if, for every number x in its domain, the number $-x$ is also in the domain AND $f(-x) = -f(x)$. The graph of an odd function is symmetric about the origin.

Neither: A function that is not even or odd is described as being "neither." It is not symmetric to the y-axis or to the origin.

Examples Determine algebraically if the functions given are odd, even, or neither.

❶ $f(x) = x^2 - 3$
$f(-x) = (-x)^2 - 3 = x^2 - 3$
$f(x) = f(-x)$ even
The function $f(x)$ is even.

❷ $g(x) = x^3 + 5$
$g(-x) = (-x)^3 + 5 = -x^3 + 5$ $-g(x) = -(x^3 + 5) = -x^3 - 5$
$g(x) = g(-x)$ not even $g(-x) = -g(x)$ not odd
The function $g(x)$ is neither.

❸ $h(x) = 2x^3 - x$
$h(-x) = 2(-x)^3 - (-x) = -2x^3 + x$ $-h(x) = -(2x^3 - x) = -2x^3 + x$
$h(x) = h(-x)$ not even $h(-x) = -h(x)$ odd
The function $h(x)$ is odd.

Building and Interpreting Functions

Algebra 2 Made Easy – Common Core Edition

Graph the function and determine if it is odd, even, or neither.

❹ $f(x) = x^2 + 2x + 1$

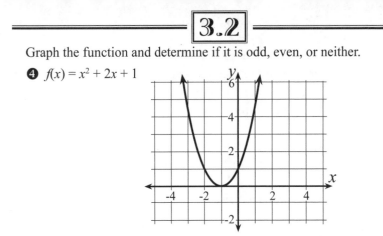

The graph is not symmetric to the y-axis or to the origin.
The function $f(x)$ is neither odd nor even.

❺ $g(x) = |-x^2|$

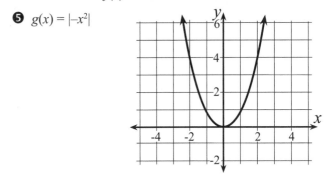

The function $g(x)$ is symmetric to the y-axis. It is an even function.

❻ $h(x) = x^3$

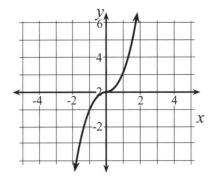

The function $h(x)$ is an odd function since its graph is symmetric to the origin.

FUNCTIONS AND TRANSFORMATIONS

REVIEW OF PARENT FUNCTIONS

Identity (Linear): $f(x) = x$ Radical: $f(x) = \sqrt{x}$

Quadratic: $f(x) = x^2$ Cubic: $f(x) = x^3$

Exponential: $f(x) = b^x$ Absolute Value: $f(x) = |x|$

RULES FOR TRANSFORMATIONS OF FUNCTIONS

$f(x)$ represents the parent function

$f(x) = x^2$ and the number 3 is used to demonstrate the rules on this page. For more parent functions and their transformations, see the next 3 pages.

$f(x) + a$: moves the graph up a units.
Ex: $f(x) + 3 = x^2 + 3$

$f(x) - a$: moves the graph down a units.
Ex: $f(x) - 3 = x^2 - 3$

$f(x + a)$: moves it to the left a units.
Ex: $f(x + 3) = (x + 3)^2$

$f(x - a)$: moves it to the right a units.
Ex: $f(x - 3) = (x - 3)^2$

$-f(x)$: reflects the graph over the x-axis.
Ex: $-f(x) = -x^2$

$f(-x)$: reflects the graph over the y-axis.
Ex: $f(x) = (-x)^2$

$a \cdot f(x)$: stretches the graph vertically (or compresses it horizontally) when $a > 1$.
Ex: $f(x) = 3x^2$

$a \cdot f(x)$: compresses the graph vertically (or stretches it horizontally) when $0 < a < 1$.
Ex: $f(x) = 0.5x^2$

Building and Interpreting Functions

Algebra 2 Made Easy – Common Core Edition   **117**

EFFECTS OF TRANSFORMATIONS ON PARENT FUNCTION GRAPHS

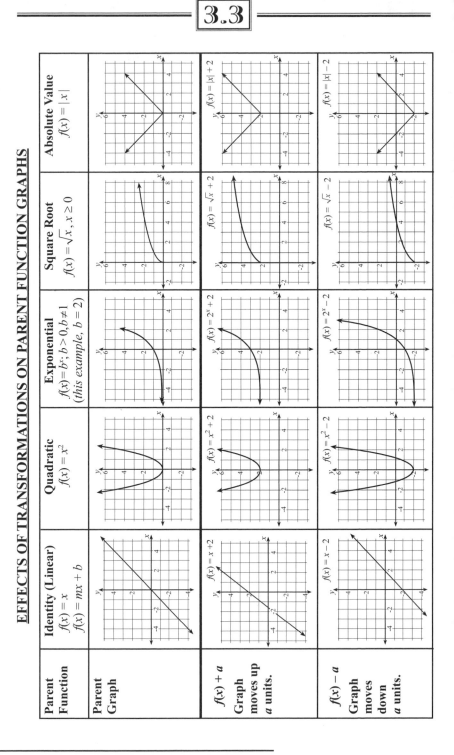

Parent Function	Identity (Linear) $f(x) = x$ $f(x) = mx + b$	Quadratic $f(x) = x^2$	Exponential $f(x) = b^x; b > 0, b \neq 1$ (this example, $b = 2$)	Square Root $f(x) = \sqrt{x}, x \geq 0$	Absolute Value $f(x) = |x|$
Parent Graph					
$f(x) + a$ Graph moves up a units.	$f(x) = x + 2$	$f(x) = x^2 + 2$	$f(x) = 2^x + 2$	$f(x) = \sqrt{x} + 2$	$f(x) = |x| + 2$
$f(x) - a$ Graph moves down a units.	$f(x) = x - 2$	$f(x) = x^2 - 2$	$f(x) = 2^x - 2$	$f(x) = \sqrt{x} - 2$	$f(x) = |x| - 2$

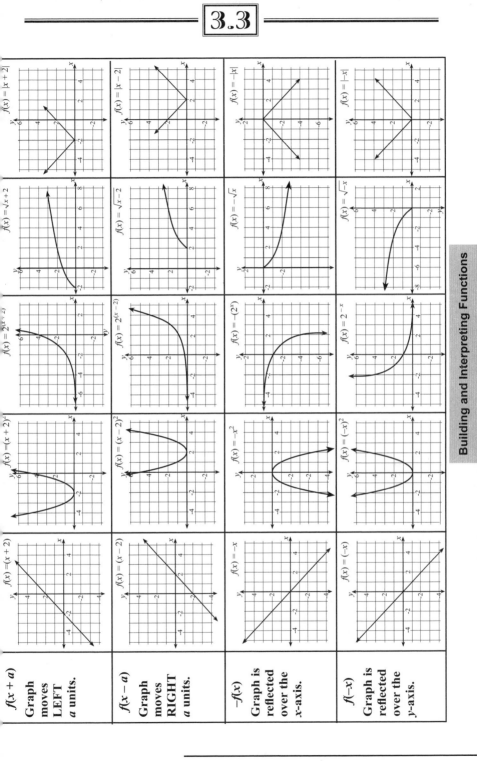

$f(x) = \|x + 2\|$	$f(x) = \|x - 2\|$	$f(x) = -\|x\|$	$f(x) = \|-x\|$
$f(x) = \sqrt{x+2}$	$f(x) = \sqrt{x-2}$	$f(x) = -\sqrt{x}$	$f(x) = \sqrt{-x}$
$f(x) = 2^{(x+2)}$	$f(x) = 2^{(x-2)}$	$f(x) = -(2^x)$	$f(x) = 2^{-x}$
$f(x) = (x+2)^2$	$f(x) = (x-2)^2$	$f(x) = -x^2$	$f(x) = (-x)^2$
$f(x) = (x+2)$	$f(x) = (x-2)$	$f(x) = -x$	$f(x) = (-x)$

$f(x + a)$	$f(x - a)$	$-f(x)$	$f(-x)$
Graph moves **LEFT** *a* units.	Graph moves **RIGHT** *a* units.	Graph is reflected over the *x*-axis.	Graph is reflected over the *y*-axis.

Building and Interpreting Functions

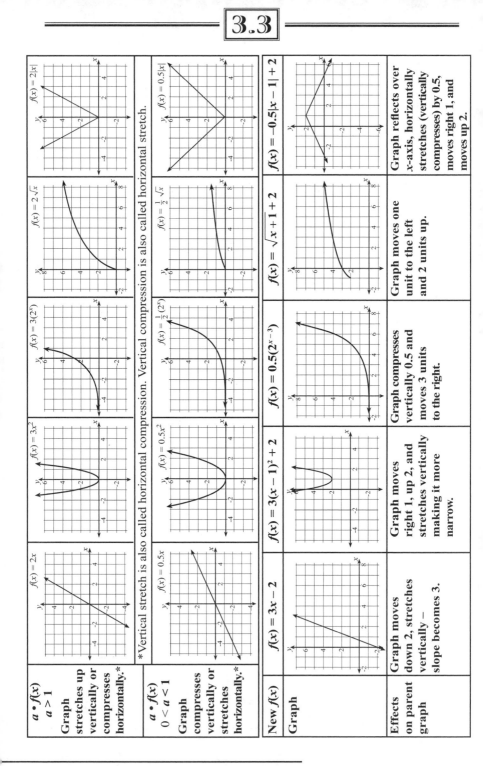

	$f(x) = 2x$	$f(x) = 3x^2$	$f(x) = 3(2^x)$	$f(x) = 2\sqrt{x}$	$f(x) = 2|x|$
$a \cdot f(x)$ $a > 1$ Graph stretches up vertically or compresses horizontally.*					
	$f(x) = 0.5x$	$f(x) = 0.5x^2$	$f(x) = \frac{1}{2}(2^x)$	$f(x) = \frac{1}{2}\sqrt{x}$	$f(x) = 0.5|x|$
$a \cdot f(x)$ $0 < a < 1$ Graph compresses vertically or stretches horizontally.*					

*Vertical stretch is also called horizontal compression. Vertical compression is also called horizontal stretch.

New $f(x)$	$f(x) = 3x - 2$	$f(x) = 3(x-1)^2 + 2$	$f(x) = 0.5(2^{x-3})$	$f(x) = \sqrt{x+1} + 2$	$f(x) = -0.5|x-1| + 2$
Graph					
Effects on parent graph	Graph moves down 2, stretches vertically – slope becomes 3.	Graph moves right 1, up 2, and stretches vertically making it more narrow.	Graph compresses vertically 0.5 and moves 3 units to the right.	Graph moves one unit to the left and 2 units up.	Graph reflects over x-axis, horizontally stretches (vertically compresses) by 0.5, moves right 1, and moves up 2.

Algebra 2 Made Easy – Common Core Edition

AVERAGE RATE OF CHANGE

Finding the average rate of change over a specific interval of a function is similar to finding the slope of a line connecting two points on a graph. When the graph is linear, we are already familiar with finding the slope, m, using the formula $m = \dfrac{y_2 - y_1}{x_2 - x_1}$. If the graph is a curve, the points involved can be connected with a line and then the slope of that line is found. The formula for the average rate of change is $Avg\ Rate = \dfrac{f(x_2) - f(x_1)}{x_2 - x_1}$. This formula can be used with a function given in algebraic form, shown as a graph, or expressed as a table.

Examples

❶ **Algebraic:** Given the function $f(x) = x^2 + 3x - 5$

  a) Find the average rate of change in the interval [2, 10].

  Solution: First evaluate $f(2)$ and $f(10)$, then substitute in the formula.
  $f(2) = 2^2 + 3(2) - 5 = 4 + 6 - 5 = 5$
  $f(10) = 10^2 + 3(10) - 5 = 100 + 30 - 5 = 125$

  $$Avg\ Rate = \frac{125 - 5}{10 - 2} = \frac{120}{8} = 15$$

  This indicates that dependent variable increases (rises) an average of 15 units for each unit of increase of the independent variable in the given interval.

  b) Using the same function, find the average rate of change in the interval [−4, −2].
  $f(-4) = (-4)^2 + 3(-4) - 5 = 16 - 12 - 5 = -1$
  $f(-2) = (-2)^2 + 3(-2) - 5 = 4 - 6 - 5 = -7$

  $$Avg\ Rate = \frac{-1 - (-7)}{-4 - (-2)} = \frac{6}{-2} = -2$$

  In the interval between $x = -4$ and $x = -2$, the function has an average rate of change of −3. The dependent variable decreases 3 units for each unit of increase in the independent variable.

Note: When graphed, part of this function is increasing and part of this function is decreasing. Therefore the average rate of change may be positive or negative, depending on the interval.

Building and Interpreting Functions

Graph: Find the average rate of change in the interval $-3 \le x \le 0$.

Solution: Locate the x values indicated and determine the value of $f(x)$. Substitute.

$$Avg\ Rate = \frac{4-(-4)}{-3-0} = \frac{8}{-3}$$

This function's dependent value, $f(x)$ or y, decreases 8 units as the independent variable, x, increases 3 units in the interval indicated.

(A line sketched between the 2 points would have a negative slope.)

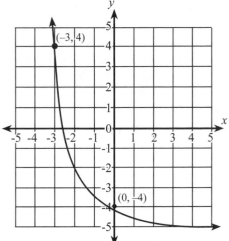

Note: This is a decreasing graph so the average rate of change between any 2 different points on the graph would be negative. However, the value of the average rate of change would be different within different intervals. For example, the average rate of change would be less between $x = 2$ and $x = 3$ than it is between $x = -3$ and $x = 0$.

Table: Given this table of values for a function find the average rate of change.

Solution: Read the corresponding values of $f(t)$ from the table and substitute.

t	$f(t)$
-4	6
-3	1
-2	-2
-1	-3
0	-2
1	1
2	6
3	13
4	22
5	33

a) Determine the average rate of change of this function from $t = -3$ to $t = 5$.

$f(-3) = 1$

$f(5) = 33$

$$Avg\ Rate = \frac{33-(1)}{5-(-3)} = \frac{32}{8} = 4$$

b) What is the average rate of change of this function from $t = -1$ to $t = 4$?

$f(-1) = -3$

$f(4) = 22$

$$Avg\ Rate = \frac{22-(-3)}{4-(-1)} = \frac{25}{5} = 5$$

Building and Interpreting Functions

Examples

❶ Lennie has recorded the average price of a gallon of gas every month for the first six months after he buys a new car. What is the average rate of change between the third month and the sixth month after he bought his car?

Month	1	2	3	4	5	6
Price in Dollars	3.29	3.59	3.39	3.09	2.99	2.67

Solution: Find the difference in the price from the 3rd to the 6th month and divide it by the number of months between them.

Conclusion: The difference in the price from month three to month six is $2.67–3.39 = $0.72.

Avg Rate of Change $= \dfrac{2.67 - 3.39}{6 - 3} = \dfrac{-0.72}{3} = -0.24$

The average rate of change in the price of gas was –$0.24 over that three month period.

❷ Shenay has been given the graph shown below that demonstrates the results of an experiment her lab partner performed. The results were recorded daily. The teacher asked Shenay to determine the average rate of change between the second day and the fifth day of the experiment.

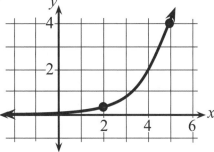

Solution: Use the formula for the average rate of change $= \dfrac{f(x_2) - f(x_1)}{x_2 - x_1}$. This is basically finding the slope of a line drawn from the two points on the graph. Read the coordinates from the graph.

Conclusion: $\dfrac{f(x_2) - f(x_1)}{x_2 - x_1} = \dfrac{4 - 0.5}{5 - 2} = \dfrac{3.5}{3} = 1.167$. The average rate of change between the first day and the 5th day of recording the results of the experiment is 1.167.

3.5

INVERSES OF FUNCTIONS

The inverse of a function is the reflection of the function over the line $y = x$. Only a one-to-one function (passes both the vertical line test and the horizontal line test) has an inverse function.

Notation: $f(x)$ is the function $f^{-1}(x)$ is the inverse

An inverse can also be named with a different function letter if it is stated that it is the inverse of the given function, such as $g(x)$ is the inverse of $f(x)$.

Find the Inverse From a Set of Ordered Pairs
Steps:
1) Determine if the points represent a one-to-one function.
2) Reverse the (x, y) to (y, x) for each pair of numbers.

Example Given a set of points in $f(x)$. Find the points of $f^{-1}(x)$

$f(x) = \{(2, 3), (3, 4), (5, 6)\}$ This is a one-to-one function.
                                        No x's are repeated, no y's are repeated.

$f^{-1}(x) = \{(3, 2),(4, 3),(6, 5)\}$ Reverse the x's and y's. These are the
                                          coordinates of the points in the inverse of $f(x)$.

Note: This is the same rule as applied when performing any reflection over the line $y = x$.

Find the Inverse Algebraically
Steps:
1) Substitute y for $f(x)$.
2) Exchange x and y in the equation.
3) Solve for y.
4) Replace y with the appropriate inverse notation.

Example Find $g(x)$, the inverse of $f(x) = 2x + 5$
$$y = 2x + 5$$
$$x = 2y + 5$$
$$\frac{x - 5}{2} = y$$
$$g(x) = \frac{x - 5}{2}$$

Algebra 2 Made Easy – Common Core Edition

Find the Inverse Algebraically and Verify

To **Verify** two functions are inverses of each other:
     1) perform and simplify the composition $f(f^{-1}(x))$.
     2) perform and simplify the composition $f^{-1}(f(x))$.
     3) if steps 1 and 2 both equal x, the functions are inverses.

Examples

❶ Find the inverse, $g(x)$, of $f(x) = 2x + 4$
and verify that $g(x)$ is the inverse of $f(x)$.

$$f(x) = 2x + 4 \qquad \text{Write down the original function}$$

$$y = 2x + 4 \qquad \text{Change } f(x) \text{ to } y$$

$$x = 2y + 4 \qquad \text{Exchange } x \text{ and } y$$

$$y = \frac{x-4}{2} \qquad \text{Solve for } y$$

$$g(x) = \frac{x-4}{2} \qquad \text{Replace } y \text{ with } g(x)$$

$$f(g(x)) = 2\left(\frac{x-4}{2}\right) + 4 = x - 4 + 4 = x$$

$$g(f(x)) = \frac{(2x+4)-4}{2} = \frac{2x}{2} = x$$

$$f(g(x)) = x \qquad g(f(x)) = x$$

$\therefore$ *They are inverses.*

❷ Prove that $f(x) = \sqrt{x+3}$ and $g(x) = x^2 - 3$ are inverses.

$$f(x) = \sqrt{x+3} \qquad and \qquad g(x) = x^2 - 3$$

$$f(g(x)) = \sqrt{(x^2 - 3) + 3} = \sqrt{x^2 - 3 + 3} = \sqrt{x^2} = x$$

$$g(f(x)) = \left(\sqrt{x+3}\right)^2 - 3 = x + 3 - 3 = x$$

They are inverses because $f(g(x)) = x$ and $g(f(x)) = x$

Building and Interpreting Functions

Graphing Inverses

Steps:

1) Graph the original function. Choose several points on the original and label them. Points can also be obtained from the Table function on the calculator.

2) Reverse the *x* and *y* coordinates of those points and graph them. This will create the points for the inverse graph which is a reflection of the original over the line $y = x$.

3) Sketch the graph and label the points.

Note: An inverse could also be graphed by finding the equation of the inverse and graphing it.

Be sure to label the graphs with $f(x)$ and $g(x)$ or as directed in the problem.

Examples

❶ Graph $f(x) = 3x - 5$ and its inverse, $g(x) = \dfrac{x+5}{3}$.

Sketch $f(x)$ and locate two points on the line. Reverse the *x* and *y* values to obtain two points on the graph of $g(x)$.

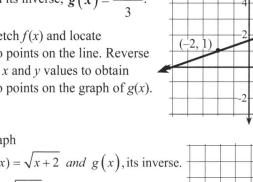

❷ Graph

$$f(x) = \sqrt{x+2} \quad and \quad g(x), \text{ its inverse.}$$

$$y = \sqrt{x+2}$$

$$x = \sqrt{y+2}$$

$$x^2 = y + 2$$

$$x^2 - 2 = y$$

$$g(x) = x^2 - 2$$

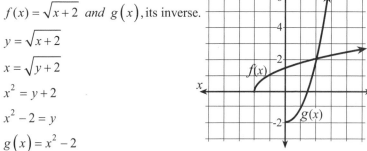

Domain and Range: The domain of $f(x)$ in this example *x* is $x \geq -2$ and the range is $y \geq 0$.

Note: The domain of the original function becomes the range of the inverse. The range of the original becomes the domain of the inverse. So although $g(x) = x^2 - 2$ is ordinarily a parabola, its domain is limited to $x \geq 0$ because the range of $f(x)$ is limited.

Algebra 2 Made Easy – Common Core Edition

EXPONENTIAL FUNCTIONS

FUNCTIONS WITH EXPONENTS

Exponential Function: A function with a variable in the exponent. The base of the exponent must be a positive number and not equal to 1. It is in the form: $y = b^x$: $b > 0$, $b \neq 1$.

The domain of an exponential function is all the real numbers. The range is $y > 0$.

Evaluate an Exponential Expression: If given the value of an exponent, apply it to the appropriate base and use the calculator to evaluate.

Examples Given: If $x = 4$, evaluate each of the following:

❶ $4^x = 4^4 = 256$

❷ $e^{2x} = e^8 = 2980.957987$ (e is on the calculator)

❸ $81^{\frac{1}{x}} = 81^{\frac{1}{4}} = 3$

Solving Equations that contain a variable with a constant exponent:
Remember that when raising a power to a power, the exponents are multiplied. This rule is used to solve equations where x is the base and it is raised to a constant exponent. (See Unit 1.1.)

Steps:

1) Isolate the variable with its exponent.

2) Raise both sides of the equation to a power that is equal to the reciprocal of the exponent. This will make the exponent on the variable = 1.

3) Solve the remaining equation

4) Check the answers.

Examples

❶ $x^{\frac{1}{2}} + 1 = 10$ *Check* :

$x^{\frac{1}{2}} = 9$ $(81)^{\frac{1}{2}} + 1 = 10$

$\left(x^{\frac{1}{2}}\right)^2 = (9)^2$ $9 + 1 = 10$

$x = 81$ $10 = 10$

Building and Interpreting Functions

② $3x^{\frac{2}{3}} = 12$ ***Check :***

$$x^{\frac{2}{3}} = 4$$

$$\left(x^{\frac{2}{3}}\right)^{\frac{3}{2}} = (4)^{\frac{3}{2}}$$

$$x = \left(\sqrt{4}\right)^3$$

$$x = 8$$

$3(8)^{\frac{2}{3}} \overset{?}{=} 12$

$3\left(\sqrt[3]{8}\right)^2 \overset{?}{=} 12$

$3(2)^2 \overset{?}{=} 12$

$12 = 12 \ \sqrt{}$

❸ $x^{-4} = 81$ ***Check :***

$$\left(x^{-4}\right)^{-\frac{1}{4}} = (81)^{-\frac{1}{4}}$$

$$x = \left(\frac{1}{81}\right)^{\frac{1}{4}}$$

$$x = \sqrt[4]{\frac{1}{81}}$$

$$x = \frac{1}{3}$$

$\left(\frac{1}{3}\right)^{-4} \overset{?}{=} 81$

$(3)^4 \overset{?}{=} 81$

$81 = 81 \sqrt{}$

❹ $x^5 = 20$ ***Check :***

$$\left(x^5\right)^{\frac{1}{5}} = (20)^{\frac{1}{5}}$$

$$x = \sqrt[5]{20} \approx 1.82$$

rounded to the 100th.
(Leave in radical form unless
directed to round.)

$(1.82)^5 \approx 20$

Due to rounding,
it is approximate.

Algebra 2 Made Easy – Common Core Edition

Exponential Expression: A constant base with a variable in the exponent.

Solving Exponential Equations: If two exponential expressions are equal to each other and they have the same bases, their exponents are equal.

Example

$$a^{2x} = a^{10+x}$$
$$2x = 10 + x$$
$$x = 10$$

If the exponential expressions are equal and they do NOT have the same base, sometimes the bases (one or both) can be rewritten as powers of the same base. If this is not possible, use logarithms. (See page 135.)

Examples

❶ $16^x = 2^8$

   $(2^4)^x = 2^8$ 16 is a power of 2.

   $2^{4x} = 2^8$ If the bases are =, the exponents are =.

   $4x = 8$

   $x = 2$

❷ $5^{2-x} = \left(\dfrac{1}{25}\right)^2$

   $5^{2-x} = \left(5^{-2}\right)^2$

   $2 - x = -4$

   $-x = -6$

   $x = 6$

❸ $4^{2x} = 32^{x-1}$

   $(2^2)^{2x} = (2^5)^{x-1}$

   $4x = 5x - 5$

   $-x = -5$

   $x = 5$

Linear, Quadratic, and Exponential Models

3.6

Exponential Growth and Decay Functions: When a given quantity is increased or decreased over time by a certain percentage we can calculate anticipated results using one of these two formulas below. (When time is the unknown, use logarithms to solve. (See page 137.)

Growth: Original amount is increasing $A_f = A_0(1 + r)^t$

Decay: Original amount is decreasing $A_f = A_0(1 - r)^t$

A_f = final amount

A_0 = original (or starting) amount

r = rate of increase or decrease for a specific time, in decimal form.

t = time (This must match the time units in the rate. If rate is per year, time must be in years. If rate is per month, or per week, time units must be months, or weeks.)

Examples

❶ Weeds are growing in Tony's front lawn at a rate of 5% per week. The lawn is 7500 square feet. If there are 40 square feet covered with weeds now, how many square feet, to the nearest integer, of the lawn will be covered with weeds after 7 weeks?

Hint: The amount of weeds at the end of 7 weeks is more than the original amount. This is an increasing function.

$A_f = ?$ $A_0 = 40$ $r = 5\% = 0.05$ $t = 7$

$A_f = A_0(1 + r)^t$

$A_f = (40)(1 + 0.05)^7$

$A_f = 40(1.05)^7$

$A_f = 56.284$; $A_f \approx 56$ *sq. feet*

---

❷ In a lab setting, insect population growth for a particular insect increases at 4.2% per day. The insect population is 5500 one week after an experiment began. What was the insect population when the experiment began? Round to the nearest integer.

Hint: The rate is per day, the time is in weeks.
Change the time to days. 1 week = 7 days

$$A_f = 5500 \quad A_0 = ? \quad r = 4.2\% = (0.042) \quad t = 1 \textit{ week} = 7 \textit{ days}$$

$$A_f = A_0(1 + r)^t$$

$$5500 = A_0(1 + 0.042)^7$$

$$5500 = A_0(1.042)^7$$

$$\frac{5500}{(1.042)^7} = A_0$$

$$A_0 = 4123.715 \text{ insects}$$

$$A_0 \approx 4124$$

❸ Charlie has $10,000 that he inherited from his grandfather. He can spend it in any way he chooses. Every year he spends 16.2% of the money in the account. If he continues to spend at this rate every year for 10 years, how much will he have left? Round to the nearest dollar.

Hint: This is a decreasing function – the amount of money at the end of 10 years is less than he started with.

$$A_f = ? \qquad A_0 = 10,000 \qquad r = 0.162 \qquad t = 10$$

$$A_f = A_0(1 - r)^t$$

$$A_f = 10,000(1 - 0.162)^{10}$$

$$A_f = 10,000(0.838)^{10}$$

$$A_f = 1707.812493$$

$$A_f \approx \$1708$$

$$\boxed{3.6}$$

SOLVING INTEREST PROBLEMS

Three types of interest that are earned on a savings account or charged for a loan are simple interest, compound interest, or continuously compounded interest.

Principal (P): The original amount of money either invested or borrowed.

Rate of Interest (r): Expressed as a decimal it is the percent that is the earned or charged over a given amount of time.

Time (t): The length of time involved in years.

Simple Interest (I): occurs when the principal is multiplied by a given rate at a specified time the amount of money is added to the original amount. This procedure can be repeated, but the calculation is always based on the principal either invested or borrowed. The formula for simple interest is $I = PRT$ where I is the interest earned at each calculation. Total interest must be added to the principal to determine the amount in the account.

Examples

❶ Melanie's Dad said he would help her save money for a concert ticket. He said he would pay her simple interest on the money she saves at a rate of 5% per month. She gave him $100 that she had saved from her part time job. The concert is scheduled for six months from now, so she has six months to save for it. How much will she be able to spend?

Solution: Determine the values of each variable that is used in the formula. The formula provides the amount of interest earned in one month and the principal stays the same, so the interest must be multiplied by 6 and added to the principal to find the final amount. The rate is based on months, and the time is based on months.

$$P = \$100.00, \quad r = .05 \text{ per month}, \quad t = 6 \text{ months}$$
$$I = Prt$$
$$I = (100)(.05)(6)$$
$$I = \$30.00$$

Conclusion: Melanie will have $130 after six months to use to purchase the ticket.

❷ What would Melanie have to pay back if she borrowed $100 from her dad and had to repay him in six months based on 5% interest per month?

Solution: The formula would be the same. Melanie would have to pay back the original $100 and pay him the $30 interest as well. She would we him $130 at the end of six months.

$$\boxed{3.6}$$

Compound Interest occurs when the principal invested at a given rate per year is compounded a specific number, n, of times per year and each time the interest is calculated, the amount of the interest is added to the present value (originally the principal). The next calculation of interest is based on the new present value in the account. In this formula, A is considered to be the "future value" of the account. P is called the present value. (At the start of the calculation, the present value is the amount of the principal.) The variable n represents the number of times per year that the interest is calculated.

$$\text{The formula is: } A = P\left(1 + \frac{r}{n}\right)^{nt}$$

Examples

❶ Juan has a summer job as a lifeguard. From his paychecks he keeps out a total of $4000 to put in the bank at the end of the summer. He finds a bank that will pay 4% annual interest compounded monthly on an account if the account is maintained for two years. If Juan leaves his money there for 2 years, how much will he have in the account? How much interest will he earn?

Solution: Notice that the rate is .04 per year and is compounded 12 times per year. The variable n in the formula will be 12. The time, t, is 2 years. P, the present value (or principal) of the account at the beginning is $4000. A will be the future value or the money in the account at the end of the given time.

Conclusion: Juan will have $4332.57 in his account at the end of 2 years. The amount of interest earned will be $332.57.

❷ If Juan takes his money out of the account in 18 months instead of 2 years, how much interest will be earned?

Solution: There isn't really enough information given in the problem to determine this. The calculation of 4% compounded monthly was based on Juan leaving the money there for 2 years. Information about the rate of interest or a penalty charged if he removes it before then is not specified. If the same interest rate was applied for 18 months then the variable t would be 1.5 and the other values would remain the same.

$$A = P\left(1 + \frac{r}{n}\right)^{nt}$$

$$A = 4000\left(1 + \frac{.04}{12}\right)^{12(1.5)}$$

$$A = 4246.92$$

Conclusion: If the rate remains the same and there is no penalty for closing the account early, the amount in the account at the end of 18 months would be $4246.92.

Linear, Quadratic, and Exponential Models

Continuous Compounding occurs when the compounding of the interest is happening continually instead of quarterly, weekly, etc. The number *e* is used in the formula instead of the $\left(1 + \frac{r}{n}\right)$ as it is the number that the expression $\left(1 + \frac{r}{n}\right)$ approaches as $n \to \infty$. The formula for continuous compounding is $A = Pe^{rt}$.

Examples

❶ George is thinking about purchasing 100 shares of a stock that sells for $12 per share. The history of the stock indicates that the stock should grow at a rate of 15% per year compounded continuously for the next 5 years. If this is an accurate prediction of the future of the stocks, what will be the value of George's stocks at the end of 5 years at which time he wants to sell it?

Solution: *P* is the amount he is investing in the stocks which is $1200. The rate is 15% per year compounded continuously and *t* is 5, the number of years he will keep the stocks.

$A = Pe^{rt}$ **Conclusion:** The value of George's stocks
$A = (1200)e^{.15(5)}$ will be $2540.40 if the past record of a 15%
$A \approx 2540.40$ increase compounded continuously continues.

❷ Suppose that June can invest $1000 in an account that will pay 11% interest compounded continuously. Which is better: For June to be given the $1000 now so she can take advantage of this investment chance, or for her to be given $1325 at the end of 3 years?

Solution: Compare the future value of $1000 invested at 11% compounded continuously for 3 years with $1325.

$A = Pe^{rt}$ **Conclusion:** June's investment would be worth
$A = (1000)e^{.11(3)}$ $1390.97 after three years. That would be better
$A \approx 1390.97$ for her to do as the other offer is for about $66
        less than that.

❸ A department store charges 1.84% per month on the unpaid balance for charge accounts. The account is compounded monthly. A customer charges $300 and pays nothing on her bill for 6 months. What is the bill at that time?

Solution: The value of *P* is $300 , *r* is .0184 compounded monthly and the time is 6 months or ½ year. 1.84% per month is 22.08% per year. In the formula, we can use the monthly rate and 6 months for the time since the rate is given per month.

$A = P\left(1 + \frac{r}{n}\right)^{nt}$ **Conclusion:** The customer will owe
$A = 300\,(1 + .0184)^{6}$ $334.68 at the end of six months.
$A = 334.68$

LOGARITHMS – ALSO KNOWN AS LOGS

Logarithm: The inverse of an exponential function. To find the inverse of a function, we exchange the x and y and then solve for y. In an exponential function, the exchange of x and y will cause y to be an exponent. In order to solve for y, we use <u>logarithms</u>. A logarithm is equal to an exponent. The form of a log function is $y = \log_b x$ where b is the base, is positive and is not equal to one. This is read, "y is the log to the base b of x." (Review for Finding Inverses see page 124.)

When writing an exponential equation in log form, the parts of the exponential equation have specific positions in the equivalent log equation.

Exponential Form: $x = b^y$ **Log form:** $\log_b x = y$

$$\boxed{\log_b x = y} \longleftarrow \text{exponent}$$

base argument

Example Find the inverse of each of the following functions.

$$f(x) = 12^x \qquad\qquad f(x) = 5^{2x}$$
$$y = 12^x \qquad\qquad y = 5^{2x}$$
$$f^{-1}: \ x = 12^y \qquad\qquad f^{-1}: \ x = 5^{2y}$$
$$y = \log_{12} x \qquad\qquad 2y = \log_5 x$$
$$f^{-1}(x) = \log_{12} x$$
$$y = \frac{\log_5 x}{2} \ \text{ or } \ y = \frac{1}{2}\log_5 x$$
$$f^{-1}(x) = \frac{\log_5 x}{2} \ \text{ or } \ f^{-1}(x) = \frac{1}{2}\log_5 x$$

Note: The example above require finding the inverse of an exponential function. Once the inverse is developed by exchanging x and y, change the new equation into a logarithm to solve for y. That is the inverse function.

Domain and Range: Since log functions are inverses of exponential functions, the domain of the log function is all the positive real numbers. The range is all real numbers.

Evaluate a Log: This means to find the exponent that must be applied to the base to equal the given number called the argument.

Examples 1, 2, and 3 contain familiar numbers. Ex. 4 requires a calculator.

❶ Evaluate $\log_2 8$.

2 is the base. What exponent must be applied to 2 to make it equal to 8? Since $2^3 = 8$, the answer is $\log_2 8 = 3$.

❷ Evaluate $\log_9 3$.

9 is the base. What exponent can be applied to 9 to make it equal to 3? Since $9^{\frac{1}{2}} = 3$, the answer is $\log_9 3 = \dfrac{1}{2}$.

❸ Evaluate $\log_4 \dfrac{1}{16}$.

4 is the base. $\dfrac{1}{4^2} = \dfrac{1}{16}$

$4^{-2} = \dfrac{1}{16}$

The solution is $\log_4 \dfrac{1}{16} = -2$.

❹ Evaluate $\log_5 2$

The base is 5 and, when raised to a power it must equal 2. This is not an answer that is familiar. Use the calculator. Change to base ten first if your calculator only uses common logs by dividing log 2 by log 5.

$\log_5 2$

$\dfrac{\log 2}{\log 5}$

$\log_5 2 \approx .4307$

Properties and Laws of Logarithms (log):

The properties and rules of exponents (See page 2.) also apply to logs. They are applied to various types of log problems in order to solve for a variable.

- Product: Add the logs. $\log_b(mn) = \log_b m + \log_b n$

- Quotient: Subtract the logs. $\log_b\left(\dfrac{m}{n}\right) = \log_b m - \log_b n$

- Power to a Power: Multiply. Since "n" is already an exponent, multiply the log (also an exponent) by "n". $\log_b(m^n) = n \cdot \log m$

Depending on the type of problem, we may need to change a single log to an expanded form, or to change an expanded log back into the form of a single log. To combine into a single log, the bases of the expanded log expression must be equal. (See Common Logs page 138.)

(*See examples on the next page.*)

Examples Expand.

❶ $\log_5(3x) = \log_5 3 + \log_5 x$

❷ $\log_3(7xy^2) = \log_3 7 + \log_3 x + 2\log_3 y$

❸ $\log\left(\dfrac{2x}{5y}\right) = \log 2 + \log x - (\log 5 + \log y)$

If no base is shown, the base is 10 and the log is called a common log.

Examples Combine into a single log.

❶ $\log_3 27 + \log_3 12 = \log_3(27 \bullet 12) = \log_3(324)$

❷ $\log x + \log(x+2) = \log[x(x+2)]$

❸ $\log x + 5\log y - 2\log z = \log\left(\dfrac{xy^5}{z^2}\right)$

Logs and Radicals: If a single log is shown in radical form, change the radical to a fractional exponent before expanding the log.

Examples Expand each logarithm.

❶ $\log\sqrt{x+3} = \log(x+3)^{\frac{1}{2}} = \dfrac{1}{2}\log(x+3)$

❷ $\log\sqrt[2]{xy^3} = \log(xy^3)^{\frac{1}{2}} = \dfrac{1}{2}(\log x + 3\log y)$ *or* $\dfrac{1}{2}\log x + \dfrac{3}{2}\log y$

Write in Log Form: Exponential equations can be put in log form, or log equations can be written in exponential form. Either form can be used when solving equations.

Examples Write in log form.

❶ $7^x = 80;$ $\log_7 80 = x$

❷ $2^x = 3;$ $\log_2 3 = x$

Examples Write in exponential form.

❶ $Log_6 x = 5;$ $6^5 = x$

❷ $Log_5 90 = x;$ $5^x = 90$

Common Logs -- Logs Using Base 10

Our decimal number system is base 10. Since that is the most common set of numbers that we deal with, the logs of those numbers are called common logs. We can work with common logs in our calculators. The absence of a base in a log expression or equation lets us know automatically that the base is 10. If another base is indicated, we can change the expression into a common log so the calculator can be used.

$Log\ x = 3$ means the same as $\log_{10} x = 3$. In exponential form it is $10^3 = x$ which can then be solved. $x = 1000$

$Log\ x = 2.57863921$ means $\log_{10} x = 2.57863921$. In exponential form this is $10^{2.57863921} = x$. This can be solved using a calculator. $x = 379$

Change of Base Formula

Logarithms in bases other than 10 can be translated to an equivalent common log using the formula: $\log_b x = \dfrac{\log x}{\log b}$. A problem with mixed log bases in it should be changed to all common logs before solving.

Examples Change the following to common logs.

❶ $\log_5 2 = \dfrac{\log 2}{\log 5} = 0.4307$

❷ $\log_2 x + \log_3 x = \dfrac{\log x}{\log 2} + \dfrac{\log x}{\log 3}$

(Examples of Common Logs on next page)

Algebra 2 Made Easy – Common Core Edition

Examples of Common Logs

Log N = 2.798325 $10^{2.798325} = N$ $\boxed{N = 628.528536}$	Evaluate Log 15 Use Log button, type in 15, Enter. Log 15 = 1.176091259	Solve & round to 4 places. $5^{2x+4} = 32$ $\log(5^{2x+4}) = \log(32)$ $(2x+4)\boldsymbol{Log}5 = \log 32$ $2x + 4 = \dfrac{\log 32}{\log 5}$
Log 10 = x $10^x = 10$ $x = 1$ $\boxed{\log 10 = 1}$	$\boxed{x = 1.176091259}$ ***Check*** Log 15 = x Means: $10^x = 15$ $10^{1.176091259} = 15$	$x = \left(\left(\dfrac{(\log 32)}{(\log 5)}\right) - 4\right) \div 2$ $x = -0.9233086048$ $\boxed{x \approx -0.9233}$ ***Check*** $5^{(2(-0.9233086048)+4)} = 32$ Be careful with the parentheses!

Solving Equations Using Logs: Use the laws of logarithms or the relationship between logarithmic and exponential form of an expression to solve log equations. Any logarithmic equation can be changed into a common log equation by changing the logarithmic equation into exponential form first or by using the change of base formula.

Note: Solving equations containing exponents and the use of logs to solve equations are closely related. Analyze each problem to find the most efficient method. Exponential equations involving e will require the use of the natural log function – abbreviated "ln" on the calculator. All properties and laws of logs in general are applied to ln.

What to do if given a value for log x: When the solution says "log x = a number", the number is the exponent of 10 that is required to find x. Use your calculator to find x.

Example   log x = 2.4759
$10^{2.4759} = x$
$x = 299.1575722$

The following examples show the use of logs in various types of equations and are rounded to a convenient place value. Be sure and follow directions about rounding when taking a test.

Examples

❶ $8^x = 50$

$x \log 8 = \log 50$

$x = \dfrac{\log 50}{\log 8}$

$x \approx 1.8813$

❷ $x^7 = 497$

$7 \log x = \log 497$

$\log x = \dfrac{\log 497}{7}$

$\log x = .385193$

$10^{.385193} = x$

$x \approx 2.43$

❸ $x^e = 40$

$e \ln x = \ln 40$

$\ln x = \dfrac{\ln 40}{e}$

$\ln x = 1.35706$

$x = e^{1.35706}$ (e is on the calculator.)

$x \approx 3.8847$

Examples 2 and 3 can be done using the reciprocal exponent. However, logs are used here to demonstrate solving with them.

❹ $\log_3 (x - 3) + \log_3 (x + 5) = 2$

Write as a single log: $\log_3 \left[(x - 3)(x + 5) \right] = 2$

Change to exponent form: $(x - 3)(x + 5) = 3^2$

Solve : $x^2 + 2x - 15 = 9$

$x^2 + 2x - 24 = 0$

$(x - 4)(x + 6) = 0$

$x = 4$ and $x = -6$ *reject*

Remember the arguments, $(x - 3)$ and $(x + 5)$, must be > 0.

$$\boxed{3.7}$$

Exponential Growth and Decay and Logs: Although some exponential growth or decay problems can be solved without logs, it is often easier to use logs – and sometimes necessary.

Remember the formulas: $A_f = A_0(1+r)^t$ and $A_f = A_0(1-r)^t$ (See page 130.)

Examples

❶ Emily deposited $5000 in an account at 4% interest. She now has $6300. How many years was the money in the account? Round to the nearest 10th.

$$A_f = 6300 \qquad A_0 = 5000 \qquad r = 4\% \qquad t = ?$$

$$A_f = A_0(1+r)^t$$

$$6300 = 5000(1+0.04)^t$$

$$\log 6300 = \log 5000 + t\log(1.04)$$

$$\log 6300 - \log 5000 = t\log(1.04)$$

$$\frac{\log 6300 - \log 5000}{\log(1.04)} = t$$

$$t = 5.892; \quad t \approx 5.9 \text{ years}$$

❷ Find the rate needed to increase $200 to $425 in 7 years.

$$A_f = 425 \qquad A_0 = 200 \qquad r = ? \qquad t = 7$$

$$425 = 200(1+r)^7$$

$$\log 425 = \log 200 + 7\log(1+r)$$

$$\log 425 - \log 200 = 7\log(1+r)$$

$$\frac{\log 425 - \log 200}{7} = \log(1+r)$$

$$\log(1+r) = 0.0467655621$$

$$10^{0.0467655621} = 1+r$$

$$1.11369 = 1+r$$

$$r = 0.11369 \text{ or } \approx 11.4\%$$

Linear, Quadratic, and Exponential Models

3.8

GRAPHING EXPONENTIAL AND LOGARITHMIC FUNCTIONS

Graphing an Exponential Function: $y = b^x$

Domain: all the reals   **Range:** all the positive reals or $y > 0$.

It is a one-to-one function.

The graph has the basic shape of a curve. Two types of graphs are possible depending on whether $0 < b < 1$ *or* $b > 1$. Both graphs intersect the y-axis at $y = 1$.

- **$b > 1$:** The graph is an increasing graph as it is read from left to right. y increases as x increases. It does not intersect the x-axis. y is always positive.

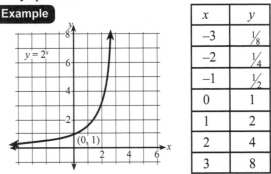

x	y
-3	$\frac{1}{8}$
-2	$\frac{1}{4}$
-1	$\frac{1}{2}$
0	1
1	2
2	4
3	8

- **$0 < b < 1$:** The graph is decreasing as it is read from left to right. y decreases as x increases. y is always positive and the curve does not intersect the x-axis.

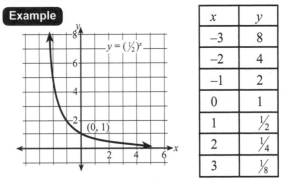

x	y
-3	8
-2	4
-1	2
0	1
1	$\frac{1}{2}$
2	$\frac{1}{4}$
3	$\frac{1}{8}$

Note: Since a negative exponent indicates that the reciprocal of the base is to be used, a negative exponent used when $b > 1$ will result in a decreasing graph. $b^{-x} = \left(\dfrac{1}{b}\right)^x$: $2^{-x} = \left(\dfrac{1}{2}\right)^x$

Algebra 2 Made Easy – Common Core Edition

Intersecting the y-axis at $(0, 1)$ is a basic characteristic of the exponential function graph. This is because any base (except 0) raised to the zero power is equal to one. For all values of b except 0, $b^0 = 1$.

Base e: e is a number that is used in many mathematical calculations and is called the natural base. It can be used as the base of an exponential function. For example, $y = e^x$. It is an irrational number and is approximately equal to 2.718. In the calculator, e is available to use in calculations. The graphs involving e have the same characteristics and rules as the graphs of exponential functions in general.

Example

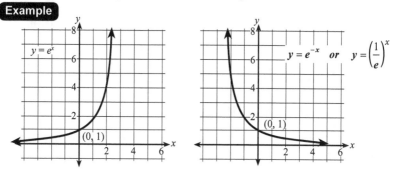

Graphing a Log Function: $\log_b x = y$ Since a logarithmic function is the inverse of the corresponding exponential function, the graph of the log function is a reflection of the exponential function over the line $y = x$.

Domain: Positive real numbers. $x > 0$ *Range:* All the real numbers.

It is a one-to-one function.

The graph has the basic shape of a curve. Two types of graphs are possible depending on whether $0 < b < 1$ *or* $b > 1$. Both graphs intersect the x-axis at $x = 1$.

Example

$y = \log_2 x; \ b > 1$ $y = \log_{0.5} x; \ 0 < b < 1$

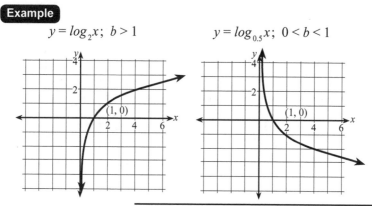

Linear, Quadratic, and Exponential Models

Natural Logs: When the natural base, e, is used in an exponential function, there is a special way to write the logarithmic function. The abbreviation for *natural log*, *ln*, is used instead of log. $\log_e x = y$ is written $\ln x = y$.

Examples

❶ Evaluate $\ln 5 = x$
  ($\ln 5 = x$ means $e^x = 5$)
  Use the "*ln*" button on the calculator to evaluate.
  $x = 1.609437912$

❷ If $\ln x = 7$, find the value of x
  Since e is the base here, simply put e^7 in the calculator.
  $x = 1096.633158$

❸ Evaluate $\ln e = x$ [$\ln e = x$ is equivalent to $\log_e e = x$]
  $\therefore e^x = e$
  $x = 1$ The *ln* of e is 1.

  (Remember, in common logs, log 10 = 1. In natural logs, $\ln e = 1$ because $\log_b b = 1$)

❹ If $y = e^x$ is the exponential function, its inverse, the log function, is written $y = \ln x$ instead of $\log_e x = y$. This is an increasing graph.

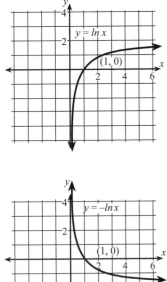

❺ If $y = e^{-x}$ is the exponential function, its inverse is written as $y = -\ln x$. This is a decreasing graph.

  Notice again, both graphs go through the point (1, 0).

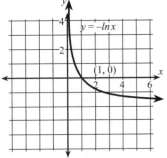

3.9

POLYNOMIAL FUNCTIONS

The most familiar polynomial functions are quadratic functions. They are in the form $f(x) = ax^2 + bx + c, a \neq 0$, which is called standard form. Standard form is needed in order to facilitate factoring, graphing, and the use of the quadratic formula.

Functions of higher degrees, with larger exponents, are considered to be in **standard form** when the variable with the highest exponent is listed first, followed by that variable with decreasing exponents until we get to the constant. The highest exponent is referred to as n, and each decreasing exponent is $n-1$, $n-2$, etc. When we get to $n-n$, that equals an exponent of zero, $n^0 = 1$.

Examples ❶ $f(x) = x^4 - x^3 + x^2 - x + 1$ ❷ $g(x) = 3x^5 + 2x^4 - 5x^3 - x^2 + 5$

Quadratic Functions: A quadratic function has the shape of a parabola when graphed. If the coefficient of the x^2 term, a, is positive the graph looks like a U, opening upward. If a is negative the graph opens downward and looks like this: ∩.

Vertex: The turning point of the parabola. It is either the minimum point when a is positive, or the maximum point when a is negative. The coordinates of the turning point can be determined algebraically by finding the equation for the axis of symmetry and using that value for the x coordinate of the vertex. Substitute it into the equation itself to find the y value, or find both values using the graphing calculator.

Maximum/Minimum Values: The y value of the vertex is either the maximum point of the graph if a is negative on the graph or the minimum point if a is positive.

Axis of Symmetry: The vertical line that goes through the vertex. Its equation is written: x = any number. It can be found algebraically by using the formula $x = \dfrac{-b}{2a}$. This provides not only the equation for the axis of symmetry, but also the x value of the vertex.

Roots, Solutions, Zeros: These terms all refer to the values of x at which $f(x) = 0$. On the graph, they are the point(s) where the graph intersects the x-axis, where $y = 0$. A parabola may have two roots, one root, or no real root. Algebraically they can be found by factoring and applying the zero product property or by using the quadratic formula to solve the equation. Real roots can be found using a graphing calculator.

Note: The roots of a function are the same if the function is changed from having a positive value of a to a negative value of a. The vertex is different and the graph opens downward instead of upward, but the roots are unchanged. The original graph is reflected over the x-axis.

Linear, Quadratic, and Exponential Models

Higher Exponent Polynomials: The rules for transformations apply to these functions as well when they are graphed.

- Adding or subtracting to the function itself (the end of the equation) moves it up or down. + moves the graph up, − moves it down.

- Adding or subtracting from *x*, within the parenthesis, moves the graph left or right. Remember a negative number moves it to the RIGHT, a positive number moves it to the LEFT.

- Multiplying the function by a number greater than 1 stretches it vertically. Multiplying by a fraction between 0 and 1 compresses the function vertically.

- Making the function negative reflects it over the *x*-axis.

COMPOSITION OF FUNCTIONS

Composition of Functions (or Composing Functions): One function is substituted into another in place of the variable. This can involve numeric substitutions or substitutions of an algebraic expression in the function in place of the variable.

Notation: $f(g(x))$ *or* $f \circ g(x)$

Composition of functions is performed by substituting the second function into the first. Put a ( ) in place of the variable in the first function. Then substitute the second function (or its result) into the parenthesis and evaluate or simplify. Always put the second function in the composition into the first function.

Examples of Numeric Compositions

$$f(x) = x + 9$$
$$g(x) = 2x + 3$$

❶ **Find:** $f(g(3))$

- Start by substituting $(2x + 3)$ for *x* in $f(x)$. Simplify, then substitute 3 for *x* and evaluate.

$f(g(3)) = (2x + 3) + 9 = 2x + 12 = 2(3) + 12 = 18$

- Alternately: Find $g(3)$ first. Then substitute that answer into $f(x)$

$g(x) = 2x + 3$
$g(3) = 2(3) + 3 = 9$
$f(x) = x + 9$
$f(g(3)) = 9 + 9 = 18$

❷ **Find:** $g(f(10)) = 2((10) + 9) + 3 = 2(19) + 3 = (38) + 3 = 41$

❸ **Find:** $g(g(5)) = 2(2(5) + 3) + 3 = 2(13) + 3 = 29$

Algebra 2 Made Easy – Common Core Edition

The "Rule" of the Composition: Here we have no number to substitute, we are combining two functions into one, producing a new equation that is equivalent to the combination of the two original functions. Substitute the second function into the place of the variable in the first function.

Examples Given: $f(x) = x - 7$ and $g(x) = x^2$

❶ $f \circ g(x) = (x^2) - 7 = x^2 - 7$ is the rule of the composition $f \circ g(x)$

❷ $f(f(x)) = (x - 7) - 7 = x - 14$ is the rule of $f(f(x))$

❸ $g \circ f(x) = (x - 7)^2 = x^2 - 14x + 49$ is the rule of $g \circ f(x)$

PERFORMING OPERATIONS WITH FUNCTIONS

Functions are added, subtracted, multiplied and divided using the following notation.

RULES:

1) $(f + g)(x)$ means add $f(x)$ and $g(x)$

2) $(f - g)(x)$ means subtract $g(x)$ from $f(x)$

3) $(f \cdot g)(x)$ means multiply $f(x)$ and $g(x)$. Be careful not to confuse this with $f \circ g(x)$ which indicates a composition.

4) $\left(\dfrac{f}{g}\right)(x)$ means divide $f(x)$ by $g(x)$

Substitute the function itself in place of $f(x)$ or $g(x)$. If given a number or a different variable expression to use, substitute that in place of x in each function and then perform the calculations.

When finding the value of the sum, difference, product, or quotient of two functions, two different methods are correct:

- Perform the operation on the functions first, then substitute the given value and evaluate.

  or

- Substitute the given value in each function, evaluate then perform the operation. These functions are the ones we'll use for examples.

Linear, Quadratic, and Exponential Models

$$\boxed{3.9}$$

Examples Given: $f(x) = x^2 + x - 2$ and $g(x) = 3x$

❶ Addition: $(f+g)(x) = f(x) + g(x)$
$$= x^2 + x - 2 + 3x$$
$$(f+g)(x) = x^2 + 4x - 2$$

Addition with Substitution

• **Substitute, then add:**

> *Find:* $(f+g)(5)$
> $[(5)^2 + 5 - 2] + 3(5) = 43$

• **Add, then substitute:**

> *Find:* $(f+g)(x+4)$
> $f(x) + g(x) = x^2 + x - 2 + 3x = x^2 + 4x - 2$
> $(x+4)^2 + 4(x+4) - 2 = x^2 + 8x + 16 + 4x + 16 - 2$
> $(f+g)(x+4) = x^2 + 12x + 30$

❷ Subtraction: $(f-g)(x) = f(x) - g(x)$
$$= x^2 + x - 2 - 3x$$
$$= x^2 - 2x - 2$$

Subtraction with Substitution

• **Substitute, then subtract:**

> *Find:* $(f-g)(5)$
> $((5)^2 + 5 - 2) - (3(5)) = 13$

• **Subtract, then substitute:**

> *Find:* $(f-g)(x+4)$
> $f(x) - g(x) = x^2 + x - 2 - 3x = x^2 - 2x - 2$
> $(x+4)^2 - 2(x+4) - 2 = x^2 + 8x + 16 - 2x - 8 - 2$
> $(f-g)(x+4) = x^2 + 6x + 6$

❸ **Multiplication:** $(f \cdot g)(x) = (f(x)) \cdot (g(x))$
$$= (x^2 + x - 2)(3x)$$
$$(f \cdot g)(x) = 3x^3 + 3x^2 - 6x$$

Multiplication with Substitution

• **Substitute, then multiply:**

Find: $(f \cdot g)(5)$

$(5^2 + 5 - 2)(3 \cdot 5) = (28)(15) = 420$

• **Multiply, then substitute:**

Find: $(f \cdot g)(x + 4)$

$f(x) \cdot g(x) = (x^2 + x - 2)(3x) = 3x^3 + 3x^2 - 6x$

$3(x + 4)^3 + 3(x + 4)^2 - 6(x + 4)$

$3(x^3 + 12x^2 + 48x + 64) + 3(x^2 + 8x + 16) - 6x - 24$

$3x^3 + 36x^2 + 144x + 192 + 3x^2 + 24x + 48 - 6x - 24$

$(f \cdot g)(x + 4) = 3x^3 + 39x^2 + 162x + 216$

❹ **Division:** $\left(\dfrac{f}{g}\right)(x) = \dfrac{f(x)}{g(x)}$

$$= \frac{x^2 + x - 2}{3x}$$

Division with Substitution

• **Substitute, then divide:**

Find : $\left(\dfrac{f}{g}\right)(5)$

$$\frac{5^2 + 5 - 2}{3(5)} = \frac{28}{15}$$

• **Divide, then substitute:**

Find : $\left(\dfrac{f}{g}\right)(x + 4)$

$f(x) \div g(x) = \dfrac{x^2 + x - 2}{3x}$ *cannot be simplified*

$$\frac{(x + 4)^2 + (x + 4) - 2}{3(x + 4)}$$

$$\left(\frac{f}{g}\right)(x + 4) = \frac{x^2 + 9x + 18}{3x + 12}$$

Linear, Quadratic, and Exponential Models

3.10

COMPARING FUNCTIONS

The graph, table of values, equation, or verbal description of a functions can be used to compare it with another function.

Examples

❶ Determine which of the following has the *smallest* maximum value?

1) $y = -|-x^2|$

2) $y = -2x^2 + 3x - 3$

3)

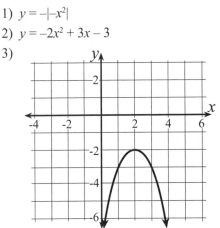

4)

x	$F(x)$
-3	6
-2	1
-1	-2
0	-3
1	-2
2	1
3	6

Solution: Read the question carefully. Using the graphing calculator, graph choices 1 and 2. Find the maximum values of each. Choice 1 has a maximum value of $y = 0$, choice 2 has a maximum value of $y = 5$. Read the maximum from the graph for choice 3. It is $y = -2$. Choice 4 has a minimum value, not a maximum.

Conclusion: The graph shown in choice 3 has a maximum value of $y = -2$ which is smaller than $y = 0$ or $y = 5$. Choice 3 is the correct answer.

Algebra 2 Made Easy – Common Core Edition

❷ Which of the following has a larger rate of change from $x = -2$ to $x = 1$?

1) $f(x) = -x^2 + 3x - 3$ 2) $f(x) = x^3$

3)

x	$f(x)$
-3	6
-2	1
-1	-2
0	-3
1	-2
2	1
3	6

4)

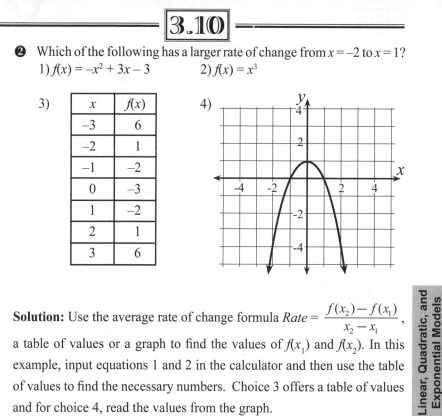

Solution: Use the average rate of change formula $Rate = \dfrac{f(x_2) - f(x_1)}{x_2 - x_1}$, a table of values or a graph to find the values of $f(x_1)$ and $f(x_2)$. In this example, input equations 1 and 2 in the calculator and then use the table of values to find the necessary numbers. Choice 3 offers a table of values and for choice 4, read the values from the graph.

Rate 1: $R_1 = \dfrac{f(1) - f(-2)}{1 - (-2)} = \dfrac{-1 - (-13)}{3} = \dfrac{12}{3} = 4$

Rate 2: $R_2 = \dfrac{f(1) - f(-2)}{1 - (-2)} = \dfrac{1 - (-8)}{3} = \dfrac{9}{3} = 3$

Rate 3: $R_3 = \dfrac{f(1) - f(-2)}{1 - (-2)} = \dfrac{-2 - (1)}{3} = \dfrac{-3}{3} = -1$

Rate 4: $R_4 = \dfrac{f(1) - f(-2)}{1 - (-2)} = \dfrac{0 - (-3)}{3} = 1$

Conclusion: The largest rate of change from $x = -2$ to $x = 1$ for these functions is choice 1. The average rate of change is 4.

Linear, Quadratic, and Exponential Models

SEQUENCES AND SERIES

Sequence: A list of terms or elements in order. The terms are identified using positive integers as subscripts of a: a_1 a_2 a_3 ... a_n. The domain is the set of consecutive positive integers starting with 1. The terms in a sequence can form a pattern or they can be random.

Series: The sum of the terms of a sequence.

Term or Element: a number in a sequence.

Subscripts: Consecutive counting (natural) number subscripts, starting with 1, that are used to identify the location of the terms in the sequence. Each term is referenced by that subscript which is called the index of the term.

Example a_3 indicates the 3rd term in the sequence.

Domain: In a finite sequence the domain is a subset of the positive integers (counting or natural numbers). In an infinite sequence the domain is the set of all positive integers

Range: The terms listed in the sequence form the range of the sequence.

Finite Sequence: Contains a specific number of terms that can be counted. The domain is a subset of the counting or natural numbers.

Examples

Sequence: 2, 5, 8, 11, 14, ... 72 *Sequence:* 5, 10, 20, 40, 80, ... 640
Index: $a_1, a_2, a_3, a_4, a_5, ... a_n$ *Index:* $a_1, a_2, a_3, a_4, a_5, ... a_n$

Infinite Sequence: Contains an unlimited number of terms that cannot be counted. The domain is the set of all natural numbers.

Examples

Sequence: 2, 5, 7, 12, 19, 24, 26, ... *Sequence:* 2, 4, 8, 16, 32, ...
Index: $a_1, a_2, a_3, a_4, a_5, a_6, a_7, ...$ *Index:* $a_1, a_2, a_3, a_4, a_5, ...$

Sequences with Patterns: An equation, also called a formula or a definition, can be used to work with a sequence. In some formulas specific numeric terms are not given. The formula depends on the index of the term desired – its location in the sequence. This type of formula is sometimes called an explicit formula or definition. A specific term can be found by substituting the number of the term for *n*. A recursive formula can be used when one or more of the terms of the sequence are given. In a recursive formula, a specific term is found by using terms located before it in the sequence.

Algebra 2 Made Easy – Common Core Edition

SEQUENCES
Finding a Specific Term in a Sequence

Recursive Definition or Formula: In a recursive definition or formula, the first term in a sequence is given and subsequent terms are defined by the terms before it. If a_n is the term we are looking for, a_{n-1} is the term before it. To find a specific term, terms prior to it must be found.

Example Find the first 3 terms in the sequence $a_n = 3a_{n-1} + 4$ if $a_1 = 5$. In this example, the first term is $a_1 = 5$. To find the 2nd and 3rd terms, $n = 2$, and $n = 3$ need to be substituted.

$$a_1 = 5$$
$$a_2 = 3(a_1) + 4; \quad a_2 = 3(5) + 4; \quad a_2 = 19$$
$$a_3 = 3(a_2) + 4; \quad a_3 = 3(19) + 4; \quad a_3 = 61$$

The three terms are 5, 19, and 61.

To *write a recursive definition* or formula when given several terms in the sequence, it is necessary to find an expression that is developed by comparing the terms and finding the process required to change each term to the subsequent term.

Example Write a recursive definition for this sequence. $-2, 4, 16, 256, \ldots$ $a_1 = -2$. Since $4 = (-2)^2$, and $16 = 4^2$, and using the last term given to us, $256 = 16^2$, a recursive definition for this sequence could be $a_n = (a_{n-1})^2$.

Explicit Formula: If specific terms are not given, a formula, sometimes called an explicit formula, is given. It can be used by substituting the number of the term desired into the formula for n. Simplify as usual.

Examples

❶ What is the 7th term in the sequence $\quad a_n = 2n - 4$
Since we want the 7th term, $n = 7$.
Substitute 7 in place of n in the equation. $\quad a_7 = 2(7) - 4$
The 7th term in this sequence is 10. $\quad a_7 = 10$

❷ What is the 5th term of the sequence $a_n = 3^n$?
Substitute 5 for n. $\quad a_5 = 3^5$
The 5th term in this sequence is 243. $\quad a_5 = 243$

❸ What are the first 3 terms in the sequence $a_n = n^2 + 1$?
3 calculations are needed: $n = 1$, $n = 2$, and $n = 3$.
$$a_1 = 1^2 + 1 = 2$$
$$a_2 = 2^2 + 1 = 5 \quad \text{The first 3 terms are: } 2, 5, 10$$
$$a_3 = 3^2 + 1 = 10$$

Note: Terms should be in simplified whenever possible.

ARITHMETIC SEQUENCE

Each term in the sequence has a common difference, **d**, with the term preceding it. The first term is labeled a_1. The formula for finding *specific terms of an arithmetic sequence* is $a_n = a_1 + (n - 1)d$, where a_n is the term desired, a_1 is the first term in the sequence, n is the location in the sequence of the term desired, and d is the common difference.

- To find **d**, COMMON DIFFERENCE: $a_2 - a_1$, $a_3 - a_2$, etc.

  If given the first term and the value of d, the formula can be used to find other terms in the sequence.

Examples

❶ Find the 7th term of an arithmetic sequence if $a_1 = 5$ and $d = 2$.

$$a_n = a_1 + (n - 1)d$$
$$a_7 = 5 + (7 - 1)(2)$$
$$a_7 = 5 + 12$$
$$a_7 = 17$$

❷ In an arithmetic series $a_1 = 5$. Find a_{10} if $a_6 = 17$ and $a_7 = 19$.

Find d: $a_7 - a_6 = 19 - 17$; $d = 2$

Use formula: $a_n = a_1 + (n - 1)d$
$$a_{10} = 5 + (9)(2)$$
$$a_{10} = 23$$

- **Recursive Formula:** Terms in the sequence are given to establish a pattern. The *general recursive formula* for an arithmetic sequence is $a_n = a_{n-1} + d$ but we have to find d.

Example Write the formula for this sequence. {1, 4, 7, 10, 13 …}

There is a common difference, d, of 3 between each pair of consecutive terms in the sequence. Each term in the sequence is found by adding 3 to the previous term. Since the terms were given, a *recursive* formula can be developed. $a_1 = 3$. $a_n = a_{n-1} + 3$.

Find the 6th term of this sequence: $a_6 = a_5 + 3$; $a_6 = 13 + 3 = 16$.

The 6th term of this sequence is 16.

To find the 20th term of this sequence we would need the 19th term to use this formula. It would make more sense to use the explicit formula, $a_n = a_1 + (n - 1)d$.

$a_{20} = a_1 + (20 - 1)(d)$; $\qquad a_{20} = 1 + 19(3)$; $\qquad a_{20} = 58$

Algebra 2 Made Easy – Common Core Edition

GEOMETRIC SEQUENCE

The consecutive terms are developed by multiplying each term in the sequence by a common ratio, r, to obtain the next consecutive term. The terms in the sequence have a common ratio, $r = \dfrac{a_2}{a_1}$. (Some texts refer to a geometric sequence as a geometric progression.)

- To find r, the COMMON RATIO: Divide a term by the term before it. $r = \dfrac{a_2}{a_1}$; $r = \dfrac{a_4}{a_3}$ ... Any two consecutive terms in a geometric sequence will have the same common ratio.

Examples

❶ $\underset{a_1\ a_2\ a_3\ a_4}{3, 6, 12, 24 \,...}$ Each pair of terms has a ratio of 2. $\dfrac{6}{3} = 2, \dfrac{12}{6} = 2, \dfrac{24}{12} = 2.$

Each term in this sequence is found by multiplying the previous term by 2.

❷ $\underset{a_1\ a_2\ a_3}{\dfrac{27}{8}, \dfrac{9}{4}, \dfrac{3}{2} \,...}$ $r = \dfrac{\frac{9}{4}}{\frac{27}{8}} = \dfrac{2}{3}$, $r = \dfrac{\frac{3}{2}}{\frac{9}{4}} = \dfrac{2}{3}$ *Common Ratio* $r = \dfrac{2}{3}$

Each term in this sequence is multiplied by 2/3 to get the next term.

- Finding a **Specific Term of a Geometric Sequence:** Use the formula $a_n = a_1 r^{n-1}$ where n is the index of the term desired, r is the common ratio of the sequence and a_1 is the first term of the sequence. Determine the value of r first, then substitute and simplify.

Using the sequence in example 1 above, 3, 6, 12, 24, find the 12th term

$n = 12, r = 2, a_1 = 3$

$a_{12} = a_1 r^{n-1}$
$a_{12} = (3)(2^{(12-1)})$
$a_{12} = 3(2048)$
$a_{12} = 6144$

Find the 7th term in this sequence: $\dfrac{1}{2}, \dfrac{1}{4}, \dfrac{1}{8} \,...$ *First find* $r : r = \dfrac{\frac{1}{4}}{\frac{1}{2}} = \dfrac{1}{2}$

$a_7 = \dfrac{1}{2} \bullet \left(\dfrac{1}{2}\right)^6$; $a_7 = \left(\dfrac{1}{2}\right) \bullet \left(\dfrac{1}{64}\right)$; $a_7 = \dfrac{1}{128}$

The Recursive Formula for a geometric sequence is $a_n = (a_{n-1})r$ when terms are given.

Example What is the 5th term of the sequence 5, 10, 20, 40,...

$r = \dfrac{20}{10} = 2, \dfrac{40}{20} = 2$; $r = 2, a_{n-1} = 40$ 40 *is the 4th term*

$a_5 = (40)(2) = 80$

Algebra 2 Made Easy – Common Core Edition
155

Building and Interpreting Functions

SERIES – THE SUM OF THE TERMS OF A SEQUENCE

ARITHMETIC SERIES

Sum of a Finite **Arithmetic Sequence**: Use the formula $S_n = \dfrac{n(a_1 + a_n)}{2}$ where n is the number of terms in the sum, a_1 is the first term in the sum, a_n is the nth term in the sum. It may be necessary to determine if the given sequence is arithmetic and to find a specific term, a_n, before using the formula.

Example What is the sum of the first 10 terms in this sequence?
2, 4, 6, 8...?

This is an arithmetic sequence. The common difference, d, is 2. Find the 10th term first. Then use the formula for the sum of series of the first ten terms:

$$a_{10} = 2 + 9(2)$$
$$a_{10} = 20$$

$$S_{10} = \frac{10(2 + 20)}{2}$$
$$S_{10} = \frac{220}{2}$$
$$S_{10} = 110$$

GEOMETRIC SERIES

Sum of a Finite **Geometric Series**: Use the formula $S_n = \dfrac{a_1(1 - r^n)}{1 - r}$ where r is the common ratio and $r \ne 1$, n is the number of terms to be included in the sum, and a_1 is the first term in the sum.

Example What is the sum of the first 8 terms of this sequence?
4, 8, 16, 32...

This is a geometric sequence and $r = 2$. $a_1 = 4$, $n = 8$.

$$S_8 = \frac{4(1 - 2^8)}{1 - 2}$$
$$S_8 = \frac{4(1 - 256)}{-1}$$
$$S_8 = \frac{4(-255)}{-1} = 1020$$

SEQUENCE AND SERIES USING SIGMA NOTATION

To use the summation feature for the sum of a series on the calculator, we use the formula for finding a specific term and do a summation using the terms asked for in the problem. The beginning number, or term, goes under the sigma, the ending number or term is on top. The expression is to the right of sigma.

Summation: $\sum\limits_{n=1}^{4}(n+2)=(1+2)+(2+2)+(3+2)+(4+2)=18$

A number can be "passed through" sigma before the calculations or after.

$3\sum\limits_{2}^{5}x = 3(2+3+4+5)=3(14)=42$

$\quad$ ***or*** $= 3(2)+3(3)+3(4)+3(5)=42$

The sum of a series, arithmetic or geometric, can be written using sigma notation, $\sum$.

Use the formula for finding a specific term with $\sum$ and indicating all the terms needed. The calculator will find each term and then add them.

Sum of an Arithmetic Series using $\sum$ on the calculator.

$\boxed{\textbf{Example}}$ Find the sum of the first 8 terms in this arithmetic series:
$$3, 9, 15, 21$$

First find d: $d = 9 - 3 = 6.$

n is the final term in the series to be included in the sum. In this example, $n = 8$. Then use the same formula with summation notation plugging in the first term and the common difference.

$$a_n = a_1 + (n-1)d$$

$$\sum\limits_{n=1}^{8} 3+(n-1)(6) = 192$$

Sum of an Geometric Series

$\boxed{\textbf{Example}}$ What is the sum of the first 6 terms of this geometric series? $3, 15, 75, 375 \ldots$

$$n = 6, r = 5, a_1 = 3.$$

$$\sum\limits_{n=1}^{6} 3(5^{n-1}) = 11718$$

Building and Interpreting Functions

Unit 4

PROBABILITY &
STATISTICS

- Make inferences and justify conclusions from sample surveys, experiments, and observational studies.

- Understand independence and conditional probability and use them to interpret data.

- Use the rules of probability to compute probabilities of compound events in a uniform probability model.

- Summarize, represent, and interpret data on a single count or measurement variable.

- Understand and evaluated random processes underlying statistical experiments.

Statistics is a process for collecting and analyzing data in large quantities, especially for the purpose of inferring population characteristics based on a random sample from that population.

Kinds of Data Studies: The data can be collected in a variety of ways. These include:

- **Population vs Sample** – The type and number of people who participate in a statistical study can impact the validity of the study. Data collection that includes all members of a population, is called a census. If only part of a population is in the study it is called a sample. In a sample, the data can be expanded to include the whole group based on the expectation that the information gathered would apply to the group as a whole. We use samples to make conclusions about the entire population. **The problem with samples is that they may not truly represent the population.** A sample is considered *random* if the probability of selecting the sample is the same as the probability of selecting every other sample. When a sample is not random, a **bias** is introduced which may influence the study in favor of one outcome over other outcomes.
  A *good sample* must:
  1. represent the whole population.
  2. be large enough.
  3. be randomly selected to eliminate bias.

- **Survey** – used to gather large quantities of facts or opinions.
  Example Political parties call to ask people to name their favorite candidate.

- **Observation** – the observer does not have any interaction with the subjects and just examines the results of an activity.
  Example Count the number of people who are using a cell phone in a mall in a particular time frame. Count the number of people wearing sneakers at school.

- **Controlled Experiment** – two groups are studied while an experiment is performed with one of them but not the other.
  Example The value of drinking orange juice to prevent a cold is measured by seeing how many people in a group given orange juice to drink for a month get a cold vs how many get a cold in the group that did not drink orange juice.

TYPES OF VARIABLES AND DATA

Categorical Variable: Allows the identification of the group or category in which the individual is placed.

Quantitative Variable: Results are numerical which allows arithmetic to be performed on them.

Categorical Data are non-numerical. The data values are identified by type. It is also called qualitative data.

Example Eye color, kind of pet owned, or the name of a favorite TV show or movie.

Quantitative Data are numerical. The data values (or items) are measurements or counts and have meaning as numbers.

Example Grades on a test, hours watching TV, or heights of students.

Univariate: Measurements are made on only one variable per observation.

Example
- Quantitative: Ages of the students in a club.

- Categorical: Kind of car owned.

Bivariate: Measurements that show a relationship between two variables.

Example
- Quantitative: Grade level and age of the students in a school.

- Categorical: Gender and favorite T.V. shows.

COLLECTION OF DATA

Data collection must be done randomly to obtain reliable results. A sample is often used to infer the characteristics of a population.

Biased: A data set that is obtained that is likely to be influenced by something – giving a "slant" to the results.

Example
- Quantitative: To determine the average age of high school students by asking only tenth graders how old they are.

- Categorical: Standing outside Yankee Stadium and asking people coming out of the stadium to name their favorite baseball team. Most would say … Yankees!

Interpreting Categorical and Quantitive Data

Unbiased: A data set that is obtained which does not favor any one group over another.

> **Example**
> - **Quantitative:** Asking people coming out of a stadium how many pets they have.
> - **Categorical:** Asking people leaving a large grocery store what their favorite flavor of soda is.

In a **survey**, a question must be asked in a neutral way to avoid bias.

> **Example** Asking, "Do you think pretzels or potato chips are a better snack?" might get different results than asking, "Do you agree that pretzels are a better snack than potato chips?" In the second question a bias may occur because the collector of data is implying that pretzels are better by the construction of the question.

In an **observational study** the location or the choice of the group of people being observed could lead to biased data. If the collector observed people leaving a movie theater on a Saturday afternoon verses on a Saturday evening his conclusions about whether or not people take children to movies may not be accurate.

> **Example** Medical researchers want to study the weights of children. They have doctors record the weights of children when they come in for their well-visits.

Experimental data collection is often done with new medicines. One group of people is treated with a particular medicine and another group, the control group, is treated with a placebo (a treatment known to have no effect). The results are then compared between the two groups.

> **Example** The Salk polio vaccine study, in which some participants were given a new vaccine and others were given a placebo injection that contained only salt water. The researchers then determined the number of participants who contracted polio.

STATISTICAL MEASURES

Data that are collected and organized are called a distribution. Typically the center of the distribution is needed and the mean, median, or mode are used to do this.

Calculations: Most of the calculations shown in this section can be easily performed using a calculator – graphing or scientific. To improve the accuracy of your work be sure to perform each calculation twice.

Sample Data:

Grade	Frequency
93	3
88	4
83	7
78	5
73	3
68	7
63	5

Frequency Table or Chart
The column on the left lists the values of the data and the right column indicates the number of times each value appears.
In calculating information using the table, the frequency must be included. This data, arranged in a frequency table, can be used for examples of mean, mode, and median.

Mean: Add the values of the data multiplied by their frequencies and divide that sum by the total number of responses. The values must be multiplied by their frequency to find the sum of the values. The symbol used for the mean is $\bar{x}$.

$$(93 \bullet 3) + (88 \bullet 4) + (83 \bullet 7) + (78 \bullet 5) + (73 \bullet 3) + (68 \bullet 7) + (63 \bullet 5) = 2612$$
Sum of the values.

$3 + 4 + 7 + 5 + 3 + 7 + 5 = 34$ Number of values included in the data.

$\bar{x} = \dfrac{2612}{34} = 76.82$ (*rounded to nearest hundreth*)

Mode: The value that appears most often in the data. This can be determined by examining the data directly or using the frequency table. There can be one mode, two modes (bimodal), or none at all. If all values appear the same number of times, there is no mode.

> **Example** In our example, there are seven grades of 83 and also seven grades of 68. The modes are 83 and 68 and the data is "bimodal."

Median: The middle value when the data is arranged in order. Arrange the values from largest to smallest, or smallest to largest, and find the middle value. If the number of values is even, average the two middle values.

> **Example** Since there are 34 values here, the middle is between the 17th and 18th grades. The 17th grade is 78 and the 18th grade is also 78 in this example, so the median is 78. If they were not the same number, the median would be the average of the 17th and 18th grades.

Discussion: The data given in example 1 have been collected from about 34 students. It might not be accurate to use this data to predict the expected measures of a population, for example all the students in a school, because the sample is small and appears to be from one class.

Interpreting Categorical and Quantitive Data

Algebra 2 Made Easy – Common Core Edition

THE SHAPE OF A DISTRIBUTION

Normal Distribution: If a curve described above has the familiar "bell shape" and is symmetrical to a vertical line in its center, the arrangement of the data is called a normal distribution. The curve is described by by its mean (center line) and its standard deviation. Figures 1, 2 and 3 are normal distributions. The area under this curve always equals 1.

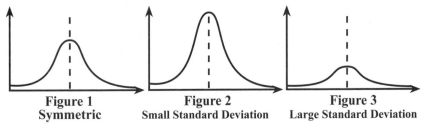

Figure 1   **Figure 2**   **Figure 3**
Symmetric   **Small Standard Deviation**   **Large Standard Deviation**

Mean: The average of the data values. It is the line of symmetry of the normal curve. The symbol for the *sample* mean is $\bar{x}$ or μ (the Greek letter mu) for the *population* mean.

$$\bar{x} = \frac{x_1 + x_2 + x_3 + ...x_n}{n}$$

Variance: The average of the squared differences the data points are from the mean. Its symbol is s^2.

$$s^2 = \frac{\sum_{i=1}^{n}(x_i - \bar{x})^2}{n}$$

Standard Deviation: A measure of the spread of the data. It is the square root of the variance. Standard deviation of a *sample* is designated by s. The symbol for standard deviation of a *population* is σ (the Greek letter sigma).

$$s = \sqrt{\frac{\sum_{i=1}^{n}(x_i - \bar{x})^2}{n - 1}} \qquad \sigma = \sqrt{\frac{\sum_{i=1}^{n}(x_i - \bar{x})^2}{n}}$$

Z-score: Tells how many standard deviations a value is above or below the mean. A z-score of one means the value is one standard deviation above the mean. A z-score can be applied to all normally distributed variables.

$$\textbf{Sample } z\text{-score} = \frac{value - mean}{standard\ deviation} \qquad \textit{or} \qquad z = \frac{x - \bar{x}}{s}$$

$$\textbf{Population } z\text{-score: } z = \frac{x - \mu}{\sigma}$$

Algebra 2 Made Easy – Common Core Edition

Data sets that commonly occur as a normal distribution are large samples of things like IQ, SAT scores, heights, weights, and standardized test scores. Inferences about population characteristics or parameters can be made using the mean and standard deviation of a normally distributed sample. Some data will form a curve when graphed, but it may not be a normal distribution curve.

Skewed Distribution: When the curve is not symmetric to the center line due to outliers in the data, the curve is described as skewed. It is not a normal distribution.

Skewed Left: Values of outliers are less than the mean. A "tail" is formed to the left of the peak. The mean is less than the median.

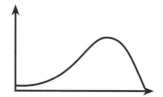

Skewed Right: Values of outliers are more than the mean. A "tail" is formed to the right of the peak. The mean is greater than the median.

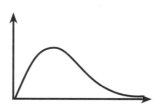

Interpreting Categorical and Quantitive Data

Characteristics of the Normal Distribution – Bell Curve

1. Mean = Median = Mode
2. A vertical line at the mean is the line of symmetry of the curve.
3. About 68% of the data is within 1 standard deviation of the mean – which includes 1 standard deviation above OR below the mean.
4. About 95% is within 2 standard deviation of the mean.
5. About 99% is within 3 standard deviation of the mean.

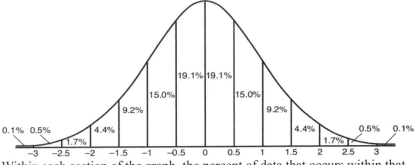

Within each section of the graph, the percent of data that occurs within that area is shown.

Within one standard deviation above the mean, 34.1 % of the data is contained. To determine the percentile of a particular piece of data, locate its position on the graph and add the percents in the sections to the left of it. Remember – percentile tells the percent of the data at or below the value given.

Note: Percentiles can also be found by adding the percents in the sections to the right and subtracting from 100.

Example SAT scores are normally distributed with an average score of 500 and a standard deviation of 100 on each section. The curve is given. Label the data values for each of the standard deviation increments along the bottom of the graph. Then answer the questions below.

Normal Curve Standard Deviation

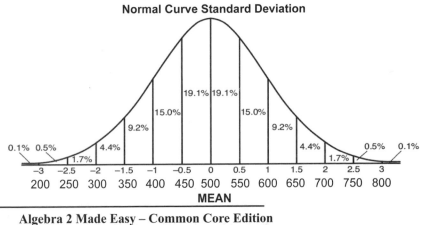

a) What score is ½ standard deviation below the mean?

– Read this from the graph: 450

b) What scores are within 1 standard deviation of the mean?

– Within 1 standard deviation of the mean includes the scores between 400 and 500, and those between 500 and 600. The solution is all scores between 400 and 600.

c) What percent of the scores fall below 1.5 standard deviation above the mean?

– There are 9 sections of the graph that are lower or to the left of 1.5 standard deviation above the mean. Add the percents for each section: 93.3%
Since 50% of the data is always to the left of the mean, you can just add 19.1 + 15.0 + 9.2 to 50 to get the answer or alternatively 100 – (4.4 + 1.7 + 0.5 + 0.1) = 93.3%

d) What is the probability that Macie scores more than 1.5 standard deviation above the mean?

– Since we just figured out that 93.3% of the data is below 1.5 standard deviation above the mean, there is 6.7% above it. The probability of scoring more than 1.5 standard deviation above the mean is 6.7%.

Interpreting Categorical and Quantitive Data

AREA UNDER THE NORMAL DISTRIBUTION CURVE
USING A CALCULATOR

The percentages indicated on the normal curve diagram indicate the percent of the area under the curve that is located between the indicated standard deviations and it is equal to the probability of an event occurring within those boundaries. When working with boundaries that are not on the standard deviation diagrams, a graphing calculator can be used to find the area under the curve within those boundaries. The information that is required includes the mean and standard deviation of the sample as well and the upper and lower boundaries of the curve.

TI Calculator Directions:

#1 Hit 2nd VARS (DISTR) and then choose 2. normalcdf(

#2 normalcdf(lowerbound, upperbound, mean, standard deviation).

normalcdf(lowerbound, upperbound, mean, standard deviation).
(Normalcdf is in the 2nd vars (DISTR) menu.)

Note: If the problem asks for "less than" or "greater than", use –99999 for negative infinity for the lower boundary, and 99999 for positive infinity for upper boundary.

Examples

❶ Given a normal distribution of values for which the mean is 70 and the standard deviation is 4.5. Find:
a) The probability that a value is between 65 and 80, inclusive.
b) The probability that a value is greater than or equal to 75.
c) The probability that a value is less than 62.

Remember:
normalcdf(lowerbound, upperbound, mean, standard deviation)

a) **The probability that a value is between 65 and 80, inclusive.**

lowerbound = 65

upperbound = 80

mean = 70

standard deviation = 4.5

The calculator should look like this

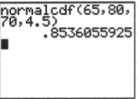

Conclusion: The area under the curve is .8536 of the total data. The probability of the value being between 65 and 80, inclusive, is 86.36%.

b) **The probability that a value is greater than or equal to 75.**

lowerbound = 75

upperbound = 99999

mean = 70

standard deviation = 4.5

Answer: 13.33%

The calculator should look like this

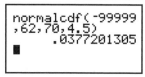

```
normalcdf(75,999
99,70,4.5)
        .1332603064
```

c) **The probability that a value is less than 62.**

lower bound = –99999

upper bound = 62

mean = 70

standard deviation = 4.5

Answer: 3.77%

The calculator should look like this

```
normalcdf(-99999
,62,70,4.5)
        .0377201305
■
```

❷ A survey indicates that for each trip to the supermarket, a shopper spends an average of 45 minutes with a standard deviation of 12 minutes in the store. The length of time spent in the store is normally distributed and is represented by the variable, *x*.

A shopper enters the store. Find the probability that the shopper will be in the store for each interval of time listed below.

a) **Between 24 and 54 minutes**

lower bound 24   mean 45

upper bound 54   standard deviation 12

Normalcdf(24, 54, 45, 12) = .73331

Answer: 73.33 %

b) **More than 39 minutes**

lower bound 39   mean 45

upper bound 99999   standard deviation 12

normalcdf(39,99999,45,12) = .69146

Answer: 69.15%

❸ **Example:** The lifetime of a battery is normally distributed with a mean life of 40 hours and a standard deviation of 1.2 hours. Find the probability that a randomly selected battery lasts longer than 42 hours.

Solution: The area under the normal curve also represents the probability.

Using the calculator, enter : Normalcdf(42,99999,40,1.2) = .0478
The probability that a randomly selected battery lasts longer than 42 hours is 4.78%.

Interpreting Categorical
and Quantitive Data

INFERENCE

Descriptive statistics is summarizing a data set based on measures of central tendency, such as the mean, and measures of spread, such as the standard deviation. **Inferential statistics** is inferring properties or characteristics about a population from a sample. This is done by using probability and sampling variability to estimate how likely it is that the sample results could have been obtained by a change given a certain population.

Statistical Inference draws conclusions about a population, based on data obtained from a sample.

Statistic: A characteristic or measure obtained by using data from a *sample*.

Parameter: A characteristic or measure obtained by using data from a *population*.

Procedure: To find the mean grade point average (GPA) of students taking class this semester at a college, choose a random sample of students from the school and record their GPAs. Suppose the statistic (the mean you get from the sample) is 2.7. Use the sample mean of 2.7 (which is a statistic) to estimate the mean for the population (which is a parameter). The best estimate of the population mean μ is the sample mean $\bar{x}$.

Is it possible that the estimate of 2.7 is the actual mean GPA for all students taking class this semester? $\bar{x}$ will often be somewhat different from the population mean μ because the sample is not a perfect representation of the population. So there is no way of knowing how close an estimate is to the population mean unless unless data is obtained from every individual in the population. This is not usually practical and is sometimes impossible. For this reason, many statisticians prefer using an estimate that contains an interval. These are called confidence intervals.

Note: A sample should be less than 10% of the population to maintain the validity of the sample mean.

CONFIDENCE INTERVALS

Even if "confidence intervals" is not familiar, the term "Margin of Error" is often used along with the results of a survey, for example, in a presidential poll. After polling 1000 eligible voters, the Star-Tribune Newspaper reported that 55% of Americans would vote for James Bean and 45% for John F Daniels +/– 3%. That plus or minus disclaimer is the **margin of error**. In other words, the margin of error means that James Bean could be favored by as much as 58 to 42 percent (55 + 3) or as low as 52 to 48 percent (55 – 3).

Either the interval contains the parameter or it does not. A **degree of confidence/confidence level** (usually a percent) can be assigned before an interval estimate is made. For example, one may wish to be 95% confident that the interval contains the true population mean. This degree of confidence/confidence level tells the likelihood that the interval estimate will contain the actual population parameter.

<u>Margin of Error</u>: The amount that a value might be above or below an observed statistic

<u>Confidence level/degree of confidence</u>: The likelihood that the interval estimate will contain the true population parameter.

The confidence level describes the uncertainty associated with a *sampling method*. Suppose we used the same sampling method to select different samples and to compute a different interval estimates for each sample. Some interval estimates would include the true population parameter and some would not. A 90% confidence level means that we would expect 90% of the interval estimates to include the population parameter; A 95% confidence level means that 95% of the intervals would include the parameter; and so on. Some people think this means there is a 90% chance that the population mean falls within the confidence interval or a 95% chance that the population mean falls within the confidence interval. It is one of the most common mistakes that people make with statistics.

<u>Confidence Interval</u>: A range or interval of values used to estimate the true value of a population parameter. Abbreviated CI.

Three confidence intervals are commonly used by statisticians—90%, 95%, and 99%. Of the three, 95% is used most frequently. It could be any confidence interval but this is the most commonly used because it creates an interval based on two standard deviations on either side of the mean.

Interpreting Categorical and Quantitive Data

Formula for the Confidence Interval of the Mean

Note: To use this formula, (1) the population standard deviation must be known, (2) the variable must either be normally distributed or the sample size must be $n > 30$, and (3) the sample is less than 10% of the population.

Formula for the confidence interval of the mean when σ is known:

$$C.I. = \overline{x} \pm z^* \, \frac{\sigma}{\sqrt{n}}$$

z^* is known as the critical value for a confidence interval. z^* changes depending on the confidence level.

Confidence Level	z^*
90%	1.645
95%	1.96
99%	2.575

Examples

❶ A sample of 50 men was taken and the mean pulse rate was $\overline{x} = 75.4$. Assume population standard deviation, σ, is 12. The pulse rates of the men are normally distributed.

a) Find the 90% confidence interval for the mean pulse rate of all men (the population mean). What is the margin of error?

Solution: Determine the value of z^* using the table. The other values needed are given. Substitute in the formula.

$$z^* = 1.645 \quad C.I. = 75.4 \pm 1.645 \left(\frac{12}{\sqrt{50}}\right)$$

Conclusion: The 90% confidence interval = 75.4 ± 1.697. This tells us that 90% of the time, the true population mean will be between 73.7 and 77.1.

b) Find the 99% confidence interval and the margin of error for the pulse rate of all men.

For a 99% confidence level: $C.I. = 75.4 \pm 2.578 \left(\frac{12}{\sqrt{50}}\right) = 75.4 \pm 4.37$

95% of the time the true population mean will be between 71.03 and 79.77. The margin of error, $E = 4.37$.

ORGANIZING CATEGORICAL DATA

TWO WAY FREQUENCY TABLES

Categorical data can be presented using a two-way frequency table. In a two-way table, there are two categorical variables involved. Within each variable there can be two or more categories included. The actual count of the data is recorded in the two-way table.

Data Set: There are 200 freshmen, 75 boys and 125 girls.
52 boys ride to school and 79 girls ride to school.
The remainder of the freshmen walk.

Example Make a two-way table and record the data.
(Give the table all necessary and appropriate titles.)

How Freshmen Get to School		
Travel Method	**Gender**	
	Boys	**Girls**
Ride	52	79
Walk	(75 – 52 =) 23	(125 – 79 =) 46

Note: For demonstration purposes, the calculations for how many of the boys and of the girls walk are shown in Figure 1. It is not necessary to show calculations on the table.

Examples

❶ What is the probability that a randomly chosen student will be a freshman girl who walks to school?

Solution: The total number of freshman is 200. The number of girls who walk to school is 46.

Conclusion: The $P(W|G) = 46/200$ or 23%

❷ What is the probability that a randomly chosen student is a boy or a walker?

Solution: $P(B) = 75/200$; $P(W) = 69/200$; $P(B$ and $W) = 23/200$
$P(B$ or $W) = 75/200 + 69/200 – 23/200 = 121/200$ or 60.5% .

See Also "Calculating Probabilities Using Two-Way Tables,"

Making Inferences and Justify Conclusions

DISPLAYING BIVARIATE DATA (2 VARIABLES)

The display of bivariate data depends on the type of data, quantitative or categorical, that is contained in the data set. Quantitative is discussed below. Categorical bivariate data can be displayed using a two-way frequency table. See page 183.

SCATTER PLOTS AND REGRESSION EQUATIONS

Quantitative Data with two variables can be displayed using a **scatter plot.** In this numerical data, the values for the ***two variables*** are paired. The plotted points often suggest a pattern (a line or a curve) which can be described using a function.

Line (or Curve) of Best Fit: It is a sketch through the points on the scatter plot such that the data points are distributed as equally as possible on both sides of the line or curve. A function (or equation) can be written to describe the line of best fit. This equation is also called a regression equation. In the example below, "Feeding the Birds", it seems that a line could be sketched through the data so the data points are equally distributed on each side of the line. The line can be defined using a linear function. (Sometimes a curve is needed.)

REGRESSION EQUATION

Regression Equation is a function that represents the graph of a line or curve of best fit. Four common examples are described here but other functions such as trigonometric functions can be used. Using a calculator to develop the scatter plot as well as the regression equation provides the most accurate information. (There more choices for regression equations depending on the data being analyzed.)

Each diagram demonstrates an example of the pattern of a scatter plot leading to a specific line of best fit for several types of regression equations. The curves of best fit are sketched in.

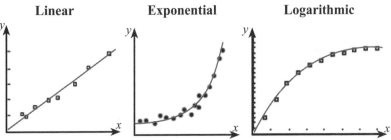

| Linear | Exponential | Logarithmic |

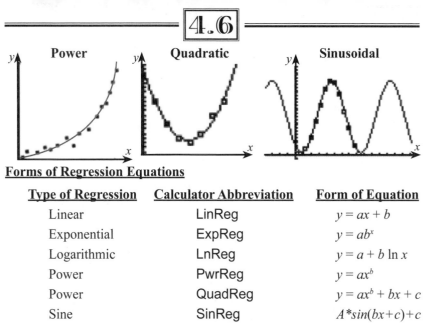

| Power | Quadratic | Sinusoidal |

Forms of Regression Equations

Type of Regression	Calculator Abbreviation	Form of Equation
Linear	LinReg	$y = ax + b$
Exponential	ExpReg	$y = ab^x$
Logarithmic	LnReg	$y = a + b \ln x$
Power	PwrReg	$y = ax^b$
Power	QuadReg	$y = ax^b + bx + c$
Sine	SinReg	$A*sin(bx+c)+c$

LINEAR REGRESSIONS

Correlation Coefficient, _r_: A number that indicates the strength and direction of a linear relationship. The regression equation gives the direction of a relationship. The strength tells how closely the data is associated with the regression equation. The value of r can be $1 \leq r \leq 1$.

Positive or Negative Value of _r_: *Refers to the upward or downward trend of the data.*

- **Positive(+)** correlation means that as x increases, y increases.
- **Negative(−)** correlation means that as x increases, y decreases.

These diagrams show a positive correlation and a negative correlation of data and a sketch of the line or curve of best fit.

Making Inferences and Justify Conclusions

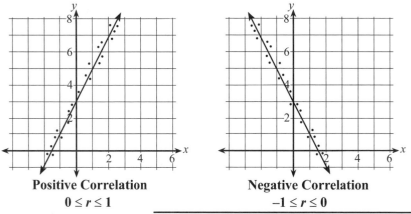

Positive Correlation	Negative Correlation
$0 \leq r \leq 1$	$-1 \leq r \leq 0$

Significance of the Numerical Value of _r_: When _r_, the correlation coefficient, is close to +1 or −1, the correlation is "strong" meaning that the regression is a close fit to the data. <u>The best regression for a set of data is the one that has a correlation coefficient closest to +1 or −1.</u>

- A strong positive correlation might be _r_ = 0.96. It indicates an upward pattern and that the data points are very close to the line or curve of best fit.

- Strong negative: _r_ = −0.96. It means a downward trend and that the data points are very close to the line of regression (or curve).

- The correlation is equally strong if _r_ = 0.96 or _r_ = −0.96

- As _r_ approaches zero, from either direction, it indicates that the regression is less closely related to the data or is "weak". If _r_ = 0.3 or _r_ = −0.3, the correlation is about the same and it is weak. The positive _r_ value indicates an upward pattern to the data, the negative _r_ indicates a downward pattern.

- A zero correlation means there is no correlation between the regression and the data. The data points are "all over the place" when compared with the line or curve of best fit.

Regression Equation: An equation that represents the line or curve of best fit.

> **Example** Several neighbors were comparing the number of bird feeders that they have in their yards in the winter with the average amount of bird seed they used in a week. A new neighbor, James, wants to start feeding the birds using 7 feeders. How much bird seed will James use in one week?

Steps:

1) Create a scatter plot to demonstrate the relationship between the number of bird feeders and the pounds of bird seed used in an average week.

2) By hand, sketch a line or curve of best fit. Determine a function to define it.

3) Describe the relationship of the independent and dependent variables.

---

4) Create a regression equation to approximately represent the line of best fit. Substitute 7 for the value of x to find the corresponding value of y.

Bird Feeders	Bird Seed (pounds)
1	3
2	4
3	9
4	10
5	15

Feeding the Birds

Solution:

a) The scatter plot for Feeding the Birds is sketched. Appropriate labels are included.

b) The data appear to cluster along a line. Sketch a line as accurately as possible between the data - some points maybe on it, some will not. After the line is sketched, it is necessary to define a function to describe this line.

c) When examining the data, as the independent variable (x) increases the dependent variable (y) also increases. This shows an upward or positive trend of the data.

d) The line of best fit can be sketched by hand. An approximate linear function, the regression equation in the form $y = mx + b$, can be created by reading the slope and y-intercept from the graph.

Substitute 7 for x. $m = 3$ $y = 3(7) - 1$
y-intercept $= -1$ $y = 20$
$y = 3x - 1$

e) Diagnostics must be on in the calculator. After putting the data into the calculator and creating the scatter plot, r is found using the "Stat Calc" key and choosing LinReg. The value of r appears in calculator screen. In this example $r \approx .97435$ which indicates a fairly strong positive correlation between the data values and the linear regression equation. As the value of x increases, the value of y also increases. The data points are closely associated with the line of best fit.

James will need about 20 pounds of bird seed for one week. (Using the calculator instead of creating the equation using the graph, a better regression equation is $y = 3x - .8$. In that case for 7 bird feeders, James will need about 20.2 lbs of bird seed. Very close to our approximation without the calculator in this example.)

Making Inferences and Justify Conclusions

NON-LINEAR REGRESSIONS

Scatter plots from bivariate data can require a curve to approximate an appropriate regression equation instead of a line. Examine the shape of the data points in the scatterplot and choose the regression equation that appears to be the best fit. See page 174 for diagrams illustrating non-linear regressions.

Examples

❶ Create a scatter plot for the area, in square feet, as measured on consecutive days. What type of regression equation appears to fit the data? Find the equation, rounded to the nearest thousandth, and discuss its accuracy. Using the rounded equation, what is the predicted area of square feet on the 12th day, to the nearest whole number?

Data	
Day	Area
0	2
1	4
2	5
3	7
4	14
5	35
6	66

Scatter Plot   **Curve of Best Fit**   **ExpReg**

Conclusion: The scatter plot of the data and the curve of best fit appear to be related to an exponential function. The area increases exponentially as the time increases. The regression equation is $y = 1.803(1.762^x)$. On the 12th day, the area should be 1615 square feet.

❷ An experiment results in the following data:

a) Create a scatter plot and sketch the line or curve of best fit for the data shown here.

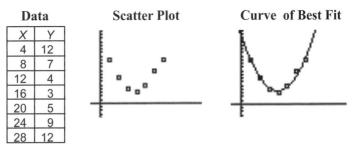

Data	
X	Y
4	12
8	7
12	4
16	3
20	5
24	9
28	12

Scatter Plot   **Curve of Best Fit**

b) Use a calculator to determine the regression equation for the line or curve of best fit.

c) Determine the value of y is x is 2.5.

d) Determine the predicted value of y if x is 50.
(Values for y may be rounded to the nearest hundredth.)

b) The scatter plot seems to be values appropriate for a quadratic equation. The line of best fit appears to be a parabola. This is called a quadratic regression, or QuadReg. Its format will be $y = ax^2 + bx + c$.

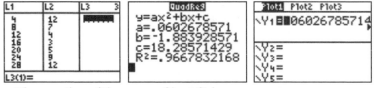

The equation of the curve of best fit is:
$y = .060267857x^2 - 1.883928571x + 18.28571429$

(*c* and *d*) Using the value function on the calculator, enter

c) $x = 2.5$; $y = 13.9525697$ *d)* $x = 100$
$\quad y = 13.95$ $\quad y = 432.57143$
$\quad y = 432.57$

❸ The average monthly temperatures, in Celsius degrees, in New York City over a thirty year period are given in the table below. January is represented by $t = 1$. Create a scatter plot and determine what type of function would best represent the curve of best fit.

Month (t)	1	2	3	4	5	6	7	8	9	10	11	12
Temp C	0.5	1.8	5.7	11.5	16.9	22.3	25.2	24.6	20.6	14.5	9.1	3.4

Source: www.weatherbase.com

Scatter Plot: The shape of the scatter plot looks like it might be a quadratic regression or it might be a sinusoidal regression. Test both and compare.

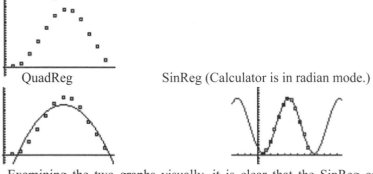

QuadReg   SinReg (Calculator is in radian mode.)

Examining the two graphs visually, it is clear that the SinReg or Sinusoidal Regression is a much better fit for the curve of best fit than the QuadReg (quadratic regression).

Conclusion: The sinusoidal regression equation is a better fit for this data. The equation is $y = 12.243 \sin(.515x - 2.187) + 12.827$ (Rounded to the nearest thousandth.)

Making Inferences and Justify Conclusions

PREDICTIONS USING REGRESSIONS

Substitute the value of x into the regression equation to find the corresponding value of y. If given y, substitute that and find x.

- **Interpolate:** Find a close approximate value that is within the given data
- **Extrapolate:** Find a close approximate value that is not within the given data – it is larger or smaller than the given data values.

Example The data in the chart has been collected in a survey. The independent variable is in the first column and the dependent variable is in the 2ⁿᵈ column.

x	y
1	7
5	20
11	32
22	50
33	66
44	80
55	93

A) Determine which regression equation fits the data best – linear, exponential, logarithmic, or power.

B) To the nearest whole number, determine the value of y when $x = 30$ and then the value of y when $x = 100$.

C) Find the value of x when $y = 100$. Round to the nearest whole number.

To solve:

A) Do all four regressions and compare the graphs to chooses the best fit. Choose that equation as your answer. Show the sketch if required.

Note: Write down all the information as you work on each regression

1) Exponential: $a = 14.18625897$, $b = 1.040918911$
   Equation: $y = 14.18625897(1.040918911)^x$

2) Logarithmic: $a = -5.292473504$, $b = 20.94195647$
   Equation: $y = -5.292473504 + 20.94195647\ ln\ x$

3) Power: $a = 6.990313513$, $b = 0.6424313887$
   Equation: $y = 6.990313513\ x^{0.6424313887}$

Solution: The power regression is the best fit.

Answer: $y = 6.990313513\ x^{0.6424313887}$

B) Do the predictions. Do not round until the last step.

$x = 30$ is within the given data. This is an interpolation. Substitute.

$$y = 6.990313513 \, (30)^{0.6424313887}$$
$$y = 62.15065544$$
$$y \approx 62$$

$x = 100$ is outside of the given data – it is an extrapolation. Use the same method.

$$y = 6.990313513 \, (100)^{0.6424313887}$$
$$y = 134.6974678$$
$$y \approx 135$$

C) $y = 100$ is an extrapolation also. Now substitute for y.

$$100 = 6.990313513 x^{0.6424313887}$$

$$\frac{100}{6.990313513} = x^{0.6424313887} \qquad \textbf{\textit{Use Logs}}$$

$$\log 100 - \log 6.990313513 = 0.6424313887 \log x$$

$$\log x = \frac{\log 100 - \log 6.990313513}{0.6424313887}$$

$$\log x = 1.798640861$$

$$10^{1.798640861} = x$$

$$x = 62.89858286$$

$$x \approx 63$$

Residual: It is the difference between the observed y value (y_o) and the predicted y value (y_p) at each observed x value of (x_o). The residuals are another way to examine the correlation of the line of best fit. Ideally the residuals should be as small as possible which would indicate that the regression line was a good fit for the data. The strength of the relationship between the data and the regression function is determined by examining the location of each plotted point compared with the line or curve of best fit.

Residual Plot: It is a scatter plot of the points representing the differences between the observed y_o, and the predicted y_p, at each observed value of x_o. The plotted ordered pairs are (x_o, y_r). The residual (y_r) is the difference between an observed y value (y_o) and the y value (y_p) predicted by the equation for the line of best fit, also called the regression line. A good fit would be indicated if most of the points of the residual plot are near the line $y = 0$. The residuals can be used to discuss the correlation of the line of best fit.

Making Inferences and Justify Conclusions

4.6

Example Determine whether the line of best fit shown in the bird feeder example on page 177 is a good fit for the data. This is called the correlation.

Plan: Use the regression function found using the calculator for best accuracy. Substitute the observed values of x, (x_o). in the function to find the corresponding predicted value of y_p. Find the difference between the predicted value of y_p, and the observed y_o value. We will call the difference, y_r. The calculation of y_r is: $y_o - y_p = y_r$.

Bird Feeders (x_o)	Bird Seed (lbs) (y_o)	$y_p = 3x_o - 0.8$	$y_o - y_p = y_r$	Residual Plot (x_o, y_r)
1	3	2.2	0.8	(1, 0.8)
2	4	5.2	−1.2	(2, −1.2)
3	9	8.2	0.8	(3, 0.8)
4	10	11.2	−1.2	(4, −1.2)
5	15	14.2	0.8	(5, 0.8)

Analysis: After plotting a graph of the residuals, we can see that most of the y points are close to $y = 0$. The observed points have small differences or residuals with respect to the line of best fit. This indicates that the line of best fit is a good fit. The data and the line of best fit have a strong correlation.

Residual Plot

Causation: Even a strong positive or negative correlation does not necessarily imply cause and effect. In the bird feeder example, the number of pounds of bird seed used does appear to be caused by how many bird feeders a person has. In other cases a strong association could be caused by other variables. Concluding that "*x causes y*" cannot be proved simply with the correlation coefficients and residuals.

Conclusion: There are many other quantitative statistical calculations that can be performed using data and their associated graphing calculator functions which will be studied in later math courses. Figure 1, on the previous page, is used to demonstrate a positive correlation. It represents an exponential regression which has a curved "line" of best fit. Other types of regression include logarithmic and power regressions, which are also both curved.

Algebra 2 Made Easy – Common Core Edition

4.6

TWO-WAY FREQUENCY TABLES

Categorical data can be presented using a two-way frequency table, there are two categorical variables involved. Within each variable there are two or more categories included. The actual count of the data is recorded in the two-way table.

Example Make a two way frequency table to record this data. Include the appropriate titles.

Data Set 1: There are 200 freshmen, 125 girls and 75 boys. 2 boys ride to school and 79 girls ride to school. The remainder of the freshmen walk.

HOW FRESHMAN GET TO SCHOOL			
Travel Method	**Gender**		**Total** (marginal row frequency)
	Boys	**Girls**	
Ride	52	79	52 + 79 = 131
Walk	23	46	23 + 46 = 69
Total (marginal column frequency)	52 + 23 = 75	79 + 46 = 125	131 + 69 = 200 & 75 + 125 = 200

Note: For demonstration purposes calculations are shown in Figure 1. It is not necessary to show calculations on the table.

What is the probability that a randomly chosen freshman is:

a) a girl?
Solution: The total number of freshmen is 200 and 125 are girls.

$$P(G) = \frac{125}{200} = 62.5\%$$

b) a walker?
Solution: Of the 200 freshman in the survey, 69 of them walk.

$$P(W) = \frac{69}{200} = 34.5\%$$

c) Is a girl who is a walker.
Solution: In the table it shows that 46 students are girls who walk.

$$P(G \text{ and } W) = \frac{46}{200} = 23\%$$

d) Is a girl or a walker?
Solution: Of the 125 girls, 46 of them are walkers. There are 69 walkers in all. "Or" means that the elements in both sets are included, however the 46 girls who walk cannot be counted twice. Add the probabilities and subtract the probability of the "overlap."

$$P(G \text{ or } W) = P(G) + P(W) - P(G \text{ and } W)$$

$$P(G \cup W) = P(G) + P(W) - P(G|W)$$

$$P(G \cup W) = \frac{125}{200} + \frac{69}{200} - \frac{46}{200} = \frac{148}{200} = 74\%$$

Making Inferences and Justify Conclusions

Algebra 2 Made Easy – Common Core Edition

PROBABILITY

Different notations are used to describe the elements or members of a set. Sometimes the elements are listed and in other cases a rule is used to define the members within the set.

DEFINITIONS:

<u>Set:</u> A group of specific items within a universe.

Example The set of integers.

<u>Subset:</u> A set whose elements are completely contained in a larger set.

Example The set of even integers is a subset of the set of integers.

<u>Complement of a Set</u>: Symbols are A', or A^C. A' contains the elements of the universal set that are not in Set A.

Example If the universe is whole numbers from 2 to 10 inclusive, and Set $A = \{2, 4, 6, 8, 10\}$, then $A' = \{3, 5, 7, 9\}$.

SYMBOLS:

$\in$ or $\subset$ means "is an element of ". $100 \in$ set of perfect squares.

$\notin$ or $\not\subset$ means "is not an element of ". $3 \notin$ set of perfect squares.

$\varnothing$ or $\{ \}$ are symbols for the "empty set" or the "null set". The empty set or null set has no elements in it.

Example If P is the set of negative numbers that are perfect squares of real numbers, then $P = \{ \ \}$ or $P = \varnothing$.

Note: Do not use $\{ \}$ and $\varnothing$ together. $\{\varnothing\}$ means the set containing the element $\varnothing$. It does not mean the empty set or null set.

<u>Mutually Exclusive:</u> Two sets that have no elements in common are called mutually exclusive sets.

<u>Notation:</u> Ways to write the elements of a set.

VENN DIAGRAMS

Symbols are used to describe relationships in set notation. A Venn Diagram can be used to demonstrate the meaning of the symbols visually. The shaded portions of the diagrams indicate the relationship between the sets as specified by the symbol.

Meaning	Set Notation Symbol	Diagram
Subset: A is a subset of B. All the elements in A are also in B.	$A \subset B$ $A \in B$	B A
Union of 2 sets is the set of elements in either A or B, or in both.	$A \cup B$	A B **Not Mutually Exclusive**
Union of 2 sets is the set of elements in both A and B.	$A \cup B$	A B **Mutually Exclusive**
Intersection of 2 sets is the set of elements that are in BOTH A and B.	$A \cap B$	A A∩B B
Complement of a set is the elements that are in the universal set but not in the given set.	A' or $\sim A$	A

Making Inferences and Justify Conclusions

Probability is the the ratio of the successful outcomes of an event to the total possible outcomes of the event.

Example What is the probability of a fair coin landing on heads if it is tossed 10 times? $P(Heads) = \dfrac{n(heads\ are\ up)}{n(total\ tosses)} = \dfrac{5}{10}$ or $\dfrac{1}{2}$ [Since it is a fair coin, it is expected to land half of the time on heads, and half on tails.]

Percents and Fractions: Probability can be shown with a fraction, a decimal, or a percent.

Probability is sometimes left in its original fraction form rather than being simplified. It may be needed in another part of the problem where a common denominator is required. It can also be written as a decimal or a percent.

Probability Values:
   $P(E)$ is never less than zero or more than one. $0 \le P(E) \le 1$
   $P(E) = 1$ when the event is sure to happen.
   $P(E) = 0$ when the event is not possible.

Complement of $P(E)$: The probability that E does NOT happen is $1 - P(E)$.
   $$P(E') = 1 - P(E)$$

Mutually Exclusive Events: Two events that have no outcomes in common.
   Example Throwing a die the lands on an even number and throwing a die that lands on an odd number. There are no even numbers that are odd.

Independent Events: Two events are independent if the outcome of one does not change the probability of the other.
   Example Spinning a spinner and tossing a coin.

Dependent Events: The outcome of one event impacts the probability of the other.
   Example Randomly choosing a red marble from a bag of marbles, not returning it, then randomly choosing a green marble from the bag.

Conditional Probability: $P(A|B)$ is read the probability of A given B. It means the probability of event A occurring after event B has occurred.

Types of Probability
Probabilities can be calculated in different ways:
   • **Theoretical probabilities:** Probabilities come from assumptions about an event and its outcomes.

   • **Empirical probabilities:** Probabilities come from data on many observations or trials.

Algebra 2 Made Easy – Common Core Edition

• **Simulation:** Probabilities that are based on data from a model. A simulation is a model in which repeated experiments are conducted to imitate a real world situation and produce similar results.

Theoretical: Based on what is expected to happen in a particular circumstance.

Example In tossing a fair die, there are six sides, equally likely to be facing up when tossed. The theoretical probability of a toss being a 5 is $\frac{1}{6}$, or 17% to the nearest percent. One side has a 5 and there are six possible outcomes.

Empirical: Based on the results of an experiment where many trials are performed and observed which then allows estimations to be made for the expected outcomes of an event happening. The more trials that are performed, the better the estimation will be.

Example A fair die is rolled 20 times and the results are recorded in the table. The empirical probability of tossing a 5 is $4/20 = 1/5 = 20\%$. During this experiment, a 5 was rolled 4 times out of the total of 20 times the die was rolled.

Number on die	Number of times rolled
1	5
2	4
3	0
4	5
5	4
6	2

Simulations: A simulation is the process of using a model to represent a real world situation instead of actually conducting the experiment. When it is very difficult or even impossible to calculate the probability of an event or conduct the actual experiment, a simulation can be used to predict the actual outcome.

Example A random number generator is used to generate a list of 50 numbers between 1 and 6 to model the rolling of a die 50 times. The results are shown in the table. The probability of rolling a 5 based on this simulation is $7/50 = 14\%$.

Number on die	Number of times rolled
1	8
2	3
3	13
4	8
5	7
6	11

From the two previous examples, the probability of rolling a 5 on a fair die is different for each type of probability. The theoretical probability is approximately 17%, the empirical is 20%, and the simulation has a probability of 14%. Empirical probabilities and probabilities based on simulations can differ from theoretical probabilities, but they are more likely to reflect the theoretical probability when the number of observations or trials is very large. The Law of Large Numbers states that a probability estimate based on a very large number of trials, tends to be closer to the true probability. If 1,000 trials were available by actual experiments or be simulation, the approximate probability would be very close to the theoretical probability.

Conditional Probability and the Rules of Probability

GENERAL PROBABILITY RULES

The type of events that are part of the experiment, the order they are performed in, and the symbols given all can have an impact on the type of calculation needed to find the probability of two or more events that occur.

INTERSECTION (AND)

And (Intersection, Notation ∩): The probability of two or more independent events occurring in a row, one and then the other, can be found by multiplying the individual probabilities. $P(A \text{ and } B) = P(A) \cdot P(B)$

Example 12% of U.S. homes own a MAC computer and 72% of U.S. homes own at least two flat screen televisions. If the two events are independent, what is the probability of owning a MAC computer and owning at least two flat screen televisions?

Solution: Let $P(M)$ represent the probability of owning a MAC computer and let $P(T)$ represent the probability of owning at least two flat screen TVs. $P(M \text{ and } T) = P(M) \cdot P(T) = (.12)(.72) = .0864$

Conclusion: The probability of owning a MAC computer and at least two flat screen TVs is approximately 9%.

UNION (OR)

Or (Union, Notation ∪): Two situations can occur when dealing with the union of two probabilities. The events can be mutually exclusive (no common or overlapping outcomes) or not mutually exclusive (some common outcomes). There is a general formula for the union that is adjusted and applied to this type of problem.

General Formula: $P(A \text{ or } B) = P(A \cup B) = P(A) + P(B) - P(A \text{ and } B)$

Mutually Exclusive: If A and B **are** mutually exclusive, then $P(A \text{ and } B) = 0$ and the formula changes to: $P(A \text{ or } B) = P(A \cup B) = P(A) + P(B)$

Examples

❶ A spinner labeled like a clock is numbered 1-12. Let A be the event the spinner lands on an even number and event B be that the spinner lands on a multiple of 3. Find $P(A \text{ or } B)$.

Solution: These two events are not mutually exclusive. There are 6 even numbers and four multiples of 3 but two of the even numbers, 6 and 12, are also multiples of 3.

$$P(A) = \frac{6}{12} \qquad P(B) = \frac{4}{12} \qquad P(A \text{ and } B) = \frac{2}{12}$$

$$P(A \text{ and } B) = P(A \cup B) = P(A) + P(B) - P(A \text{ and } B)$$

$$P(A \cup B) = \frac{6}{12} + \frac{4}{12} - \frac{2}{12} = \frac{8}{12} \text{ or } \frac{2}{3}$$

Algebra 2 Made Easy – Common Core Edition

❷ Using the same spinner, let A be the event the spinner lands on an odd number and event B be that the spinner lands on a multiple of 4. Find $P(A \cup B)$.

Solution: The events are mutually exclusive. A number cannot be odd and a multiple of 4 at the same time.

$$P(A) = \frac{6}{12} \qquad P(B) = \frac{3}{12} \qquad P(A \text{ and } B) = 0$$

$$P(A \text{ or } B) = P(A) + P(B) - P(A \text{ and } B)$$

$$P(A \text{ or } B) = \frac{6}{12} + \frac{3}{12} = \frac{9}{12} = \frac{3}{4}$$

❸ Let event A be that George tosses a fair die and gets an even number. Let event B be that Pete spins a spinner with 26 letters from the alphabet and gets a vowel. There are 6 vowels in the alphabet. Find $P(A \text{ and } B)$ to the nearest whole percent.

$$P(A) = \frac{3}{6} \qquad\qquad P(B) = \frac{6}{26}$$

Solution: $\quad P(A \text{ and } B) = P(A) \cdot P(B)$

$$P(A \text{ and } B) = \frac{3}{6} \cdot \frac{6}{26} = \frac{18}{156} \approx .1153 \approx 12\%$$

CONDITIONAL(GIVEN) $P(A \mid B)$

Conditional Probability notation is used when one event happens after a given event has already occurred. $P(B|A)$ is read, "The probability of event A occurring after event B has already occurred."

The formula for the conditional probability is:

$$P(B|A) = \frac{P(A \text{ and } B)}{P(B)} \quad or \quad \frac{P(A|B)}{P(B)} \ .$$

(See also page 195.)

VENN DIAGRAMS

Venn diagrams and Two Way Frequency Tables are often used to help visualize problems and calculate probabilities.

Examples

❶ The Sundae Funday Ice Cream Shop is testing out two new flavors, Math Brownie Explosion and Two-tti Frutti Pi. The shop conducted a survey of 150 customers which showed that 94 customers liked Math Brownie Explosion, 76 customers liked Two-tti Frutti Pi, and 34 liked both flavors, and 14 customers did not like either. Interpret and explain this situation.

a) **Create a Venn Diagram based on the information given.**

Let *M* represent liking Math Brownie Explosion and *T* represent liking Two-tti Fruitti Pi.

Two New Ice Cream Flavors

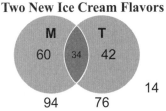

b) **Determine the probabilities of each of the following and round answers to the nearest hundredth.**

- What is the probability that one of those customers selected at random would like Math Brownie Explosion?
  Solution: $P(M) = 60/150 + 34/150 = 94/150 = 47/75 \approx .63$

- What is the probability that one of those customers selected at random would like Two-tti Frutti Pi?
  Solution: $P(T) = 42/150 + 34/150 = 76/150 = 38/75 \approx .51$

- What is the probability that one of those customers selected at random would like both Math Brownie Explosion and Two-tti Frutti Pi?

  Solution: The intersection of sets *M* and *T* consists of the elements that are in both sets *M* and *T*. It is the overlapping section of the Venn diagram. $P(M \cap T) = 34/150 = 17/75 \approx .23$

- What is the probability that one of those customers selected at random would like Math Brownie Explosion or Two-tti Frutti Pi?

  Solution: The union of sets *M* and *T* consists of all of the elements that are in either *M* or *T* or both. Subtract the elements that are in both.
  $P(M \cup T) = 94/150 + 76/150 - 34/150 = 136/150 = 68/75 \approx .91$

- What is the probability that one of those customers selected at random
  $P(M) = 1 - P(M) = 1 - 94/150 = 56/150 = 28/75 \approx .37$

Algebra 2 Made Easy – Common Core Edition

❷ A survey of 50 people found that 38 people believe in ghosts, 25 people believe in aliens, and 17 believe in both ghosts and aliens.

a) **Create a Venn Diagram and a two-way frequency table.**

Determine how many people:

Believe in ghosts?	Believe in aliens?	Like both?
38	25	17
Only believe in ghosts?	Only believe in aliens?	Like neither?
38 – 17 = 21	25 – 17 = 8	50 – (21 + 8 + 17) = 4

What Do You Believe In

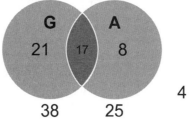

Create a two-way frequency table from the data in the Venn diagram.

	Believes in ghosts	Does not believe in ghosts	Total
Believes in Aliens	17	8	25
Does not Believe in Aliens	21	4	25
Total	38	12	50

b) **Determine the probabilities of each of the following and round answers to the nearest hundredth.**

- What is the probability that one of those people selected at random believes in ghosts?

$$P(G) = \frac{38}{50} = \frac{19}{25} = .76$$

- What is the probability that one of those people selected at random believes in aliens?

$$P(A) = \frac{25}{50} = \frac{1}{2} = .50$$

- What is the probability that one of those people selected at random believes in ghosts and believes in aliens?

$$P(G \cap A) = \frac{17}{50} = .34$$

- What is the probability that one of those customers selected at random believes in ghosts or believes in aliens?

$$P(G \cup A) = 38/50 + 25/50 - 17/50 = 46/50 = 23/25 = .92$$

Conditional Probability and the Rules of Probability

INDEPENDENT AND CONDITIONAL EVENTS

Multiple events that occur can be dependent events or independent events. A test can be used to determine if two events are independent or not.

Independent Events: Two events, A and B, that occur so that if A occurs it does not change the probability of B occurring.

Examples

❶ Tossing a 3 on a fair die, then flipping a coin that lands on tails.

❷ Choosing a king from a deck of cards, replacing it, and then choosing a queen as the second card.

Probability Involving Independent Events: The probability of two independent events occurring in sequence is the product of their individual probabilities. This is called the Multiplication Rule and multiplication is represented as "and".

If two events are independent the probability of them both occurring is: $P(A \text{ and } B) = P(A) \cdot P(B)$.

Examples

❶ Tossing a 3 on a fair die, A, and tossing a coin that lands on tails, B.

$$P(A) = \frac{1}{6} \qquad P(B) = \frac{1}{2}$$

$$P(A \text{ and } B) = \frac{1}{6} \cdot \frac{1}{2} = \frac{1}{12}$$

❷ Choosing a King from a deck of cards (A), replacing it, and then choosing a Queen from the deck (B)

$$P(A) = \frac{4}{52} \qquad P(B) = \frac{4}{52}$$

$$P(A \text{ and } B) = \frac{4}{52} \cdot \frac{4}{52} = \frac{16}{2704} = \frac{1}{169}$$

Proving 2 events are Independent: Two events are independent if the probability of the second event occurring does not affect the probability of the second event occurring.

This can be shown by either $P\left(\dfrac{B}{A}\right) = P(B)$ or $P\left(\dfrac{A}{B}\right) = P(A)$.

Examples

❶ Roll a single 6-sided die and consider the following events.
Let event O be that you roll and odd number and let event D be that you get a multiple of 3. Are the events O and D independent?

Solution: We can show this using both variations of the formula and the following probabilities:

$P(O) = \dfrac{3}{6} = \dfrac{1}{2}$ (There are three odd numbers.)

$P(D) = \dfrac{2}{6} = \dfrac{1}{3}$ (The numbers 3 and 6 are divisible by three.)

$P\left(\dfrac{O}{D}\right) = \dfrac{1}{2}$ (There is one odd number out of the two numbers divisible by 3.)

$P\left(\dfrac{D}{O}\right) = \dfrac{1}{3}$ (There is one number divisible by three out of three odd numbers.)

To determine independence, substitute the values into the formula

$$P\left(\dfrac{O}{D}\right) = P(O) \qquad \text{or} \qquad P\left(\dfrac{D}{O}\right) = P(D)$$

$$\dfrac{1}{2} = \dfrac{1}{2} \qquad\qquad\qquad \dfrac{1}{3} = \dfrac{1}{3}$$

Conclusion: Since the probabilities are the same, the events are independent.

❷ When a student is selected at random from a very large high school, the probability that the student has a smart phone is 0.68, the probability that the student has a tablet is 0.45, and the probability a student has both a smart phone and a tablet is 0.39. Let A be the event the student has a smart phone and B be the event the student has a tablet. Are these events independent?

Solution: You can also show independence by substituting values into the formula $P(A \text{ and } B) = P(A) \cdot P(B)$ and verify that they are equal.

$P(A) = 0.68 \qquad P(B) = 0.45 \qquad P(A \text{ and } B) = 0.39$

$P(A \text{ and } B) = P(A) \cdot P(B) = 0.68 \times 0.45 = 0.306$. Since this does not equal 0.39, the events are not independent.

Conditional Probability and the Rules of Probability

DEPENDENT EVENTS

When one event has an impact on the probability of the second event occurring, these are called dependent events. To find the probability of a particular outcome, multiply the probability of the first event by the probability of the second.

Examples For dependent events, we can rearrange the conditional probability formula to get $P(A \text{ and } B) = P(A) \cdot P(B|A)$.

❶ What is the probability of choosing two kings from a deck of cards where the first king drawn is not put back in the deck and then the second king is drawn.

Solution: The probability of drawing the first king (A) is $\frac{4}{52}$. After removing that king from the deck, there are 51 cards left in the deck, 3 of which are kings. The probability of choosing a second king (B) is $\frac{3}{51} \cdot P(A|B) = \frac{4}{52} \cdot \frac{3}{51} = \frac{12}{2652} = \frac{1}{221}$

❷ A bag contains 5 chocolate chip cookies, 3 sugar cookies, and 7 vanilla cookies. What is the probability of Joe picking a chocolate chip cookie from the bag and eating it, then picking a vanilla cookie?

Solution: There are 15 cookies in the bag to begin with, 5 of which are chocolate chip (A). $P(A) = \frac{5}{15}$ When he picks the next cookie (B) there are still 7 vanilla cookies in the bag but only 14 cookies altogether. $P(B) = \frac{7}{14}$. $P(A \text{ and } B) = \frac{5}{15} \cdot \frac{7}{14} = \frac{1}{3} = \frac{1}{2} = \frac{1}{6}$

CONDITIONAL PROBABILITY: $P(B|A)$

The probability that an event A occurs given that event B has already occurred. The notation for this is $P(B|A)$ and it is read, "The probability of B given A." Conditional probability is used with dependent events.

Formula: $P(B|A) = \dfrac{P(A \mid B)}{P(A)}$

Examples

❶ Find the probability that a king randomly drawn from a deck of cards is a black card. Let event A be the drawing of a king and let event B be the drawing of a black card. Find $P(B|A)$.

Solution:

- Find the probability of drawing a king from the deck. There are 4 kings in the deck of 52 cards. $P(A) = \dfrac{4}{52}$.

- Find $P(A|B)$. There are 2 black kings in the deck. $P(A|B) = \dfrac{2}{52}$.
  Note: $P(A|B) = P(B|A)$ and either one can be used.

- Substitute those values in the formula for conditional probability. Simplify.

$$P(B|A) = \frac{P(A \mid B)}{P(A)}$$

$$P(B|A) = \frac{P(A \mid B)}{P(A)} = \frac{\dfrac{2}{52}}{\dfrac{4}{52}} = \frac{2}{4} = \frac{1}{2}$$

❷ At Lincoln High School the probability that a student takes French and Music is 0.078. The probability that a student takes music is 0.53. What is the probability, to the nearest whole percent, that a student takes Music given that the student is taking French?

Solution: $P(\text{Music}|\text{French})$ is the conditional setup for this problem.

$$P(M|F) = \frac{P(F \text{ and } M)}{P(M)} = \frac{0.078}{0.53} \approx 0.1471 \approx 15\%$$

Conclusion: There is about a 15% probability that a student who is taking French is also taking Music.

Conditional Probability and the Rules of Probability

CONDITIONAL PROBABILITY AND TWO – WAY TABLES

The information needed to calculate conditional probabilities is readily available on a two-way table. It involves finding the probability for the condition of the example: $P(A)$ in $P(B|A)$ and the probability of the probability of the intersection of A and B.

Using the information in the two-way frequency table below, explain and calculate the probability of the following:

HOW FRESHMAN GET TO SCHOOL			
Travel Method	Gender		Total (marginal row frequency)
	Boys	Girls	
Ride	52	79	131
Walk	23	46	69
Total (marginal column frequency)	75	125	200

Example What is the probability that a randomly chosen freshman is:

a) A student who rides (R), given that it is a boy, (B)? Use the formula and explain another method of determining this probability.

Solution: The condition here is that the student must be a boy (B). Of the boys how many ride to school (R)? The notation for this is $P(B \mid R)$. Find $P(B)$ and $P(R \mid B)$, the intersection of boys and ride and substitute in the formula.

$$P(B \mid R) = \frac{52}{200} \qquad P(B) = \frac{75}{200}$$

$$P(R \mid B) = \frac{P(B \mid R)}{P(B)}$$

$$P(R \mid B) = \frac{\dfrac{52}{200}}{\dfrac{75}{200}} = \frac{52}{75} \approx 69\%$$

Another method of finding $P(R|B)$ is using the numbers directly from the two-way frequency table. The condition is that the student chosen is a boy. Use the column labeled Boys to find the total number of boys. The next requirement is that he is a rider. Use the row labeled Ride. The cell where Boys intersects with Ride is the numerator of the probability fraction, the total of the Boys column is the denominator.

---

b) A student who rides (R) to school given that the student is a girl (G)?

Solution: The condition of being a rider is the first occurrence in this problem. Then, from the riders, how many are girls is the next part of the probability. Find $P(R)$ and $P(G \mid R)$, the intersection of girls and ride. Show two methods of finding this conditional probability.

$$P(R) = \frac{131}{200} \qquad P(R \mid G) = \frac{79}{200}$$

$$P(G \mid R) = \frac{P(R \mid G)}{P(R)}$$

$$P(G \mid R) = \frac{\frac{79}{200}}{\frac{131}{200}} = \frac{79}{131} \approx 60\%$$

An alternate method to solve this is to use the total of the row titled Ride as the denominator and the number in the cell where the Ride row intersects with the Girls column to find the probability fraction.

PROBABILITIES USING TWO-WAY TABLES AND VENN DIAGRAMS

Two-Way Frequency Tables

Tables are often used as a tool to organize and present data. Two-way frequency tables are used to organize bivariate categorical data in rows and columns and interpret the data. Two-way frequency tables can be used to calculate probabilities.

Examples

❶ The table below shows the results of a survey in which young adults ages 18-24 were asked if they ever used Instagram (yes or no).

	Female (*F*)	Male (*M*)	TOTAL
Has Used Instagram (*Y*)	216	172	388
Has Never Used Instagram (*N*)	54	68	122
TOTAL	270	240	510

a) Use the table above to find the probability of randomly selecting a young adult who is female (*F*) and has used Instagram (*Y*) to the nearest percent.

There are 216 females who have used Instagram out of total of 510 young adults. The probability of a young adult who is female and has used Instagram is $\frac{216}{510}$. Therefore, 42% of the young adults are female and have used Instagram.

b) Use the table above to find the probability of randomly selecting a young adult who is female or has used Instagram to the nearest percent. Explain why $P(F \text{ or } Y) \neq P(F) + P(Y)$.

Solution: $P(F) = \frac{270}{510}$ 270 students are girls.

$P(Y) = \frac{388}{510}$ 388 students use Instagram.

$P(F \text{ and } Y) = \frac{216}{510}$ 216 girls use Instagram.

$P(F \text{ or } Y) = P(F) + P(Y) - P(F \text{ and } Y) = \frac{270}{510} + \frac{388}{510} - \frac{216}{510} = \frac{442}{510} = \frac{13}{15} = .86$

Conclusion: The probability of randomly selecting a young adult who is female or has used Instagram to the nearest percent is 87%.

$P(F \text{ or } Y) \neq P(F) + P(Y)$ because there was overlapping.

$P(F) + P(Y) = \frac{270}{510} + \frac{388}{510} = \frac{652}{510}$ which is not even a possible probability.

Probabilities are between 0 and 1 inclusive and 658/510 is greater than 1. This is because the overlapping was not subtracted and females (F) that have used Instagram (Y) were counted for twice.

c) Use the table above to find the probability that a randomly selected young adult who is female has used Instagram. Indicate whether or not it is a conditional probability.

Solution: There are 216 females who have used Instagram out of a total of 270 (216 + 54) females total. The probability that a female has used Instagram is $216/270 = 4/5 = 0.8$. Therefore, 80% of females have used Instagram. This represents a conditional probability because it is based on only the females.

$$P(Y \mid F) = \frac{P(F \text{ and } Y)}{P(F)} = \frac{(216/510)}{(270/510)} = \frac{216}{270}$$

Conclusion: The probability that the young adult is a female who has used instagram is 0.8 or 80%.

❷ Movie Warehouse asked 1,000 customers how they prefer to view movies. The results are summarized in the two-way table below.

	DVD or BluRay	Netflix or similar provider	Online Download
Age under 35	52	145	186
Age 35-55	108	110	56
Age over 55	250	75	18

a) What is the probability that a customer would prefer to view movies via online download given that they are under 35 years old?
   - The probability of randomly selecting a customer that would prefer to view movies via online download given that they are under 35 years old is 186/1000 or 18.6%.

b) What is the probability that a customer would prefer to view movies via DVD or BluRay given that they are between the age of 35 and 55?
   - The probability of randomly selecting a customer that would prefer to view movies via DVD or BluRay given that they are between the age of 35 and 55 is 108/1000 or 10.8%.

Calculating the probabilities for this example is very simple because the population of customers is 1000. Since it is simple working with a population of 1000, a hypothetical 1000 table is often used when only some of the information is given. In many cases, you may only know the probabilities of some of the events. In these situations, it may be possible to create a hypothetical 1000 table to use to calculate probabilities.

Conditional Probability and the Rules of Probability

Hypothetical 1000 Table: A hypothetical 1000 table is a two-way table that can be constructed using given probabilities. It represents a hypothetical population of 1000 which makes the given probabilities easier to understand and work with.

Example A survey was taken of students' genders and their preference for basketball, soccer, or track. Ten percent of girls and 8% of boys preferred track, 12% of boys preferred basketball, and 30% of girls preferred soccer. Fifty percent of the students preferred soccer. Make a two-way frequency table comparing the gender of the students and their preferences for basketball, soccer, and track with a hypothetical population of 1000 students. Use the two-way frequency table to answer the questions below.

Steps

1) Make a hypothetical 1000 table. Use the given probabilities to determine the number of students.

Multiply each probability in decimal form by 1000 since the table represents a population of 1000.

For example, 10% of girls preferred track. Multiply .10 by 1000 to find that 100 females play track.

8% of boys preferred track. Multiply .08 by 1000 to find that 80 males play track.

	Basketball	Soccer	Track	Totals
Male	$(.12)(1000) = 120$	$500 - 300 = 200$	$(.08)(1000) = 80$	$120 + 200 + 80 = 400$
Female	$320 - 120 = 200$	$(.3)(1000) = 300$	$(.1)(1000) = 100$	$200 + 300 + 100 = 600$
Total	$1000 - (500 + 180) = 320$	$(.5)(1000) = 500$	$80 + 100 = 180$	1000

2) Use the information that you have filled in to complete the table.

Note: For demonstration purposes, the calculations used to complete the table are shown above. It is not necessary to show calculations in the table.

	Basketball	Soccer	Track	Totals
Male	120	200	80	400
Female	200	300	100	600
Total	320	500	180	1000

a) What is the probability that a student preferred soccer given that the student was a girl? $P(B|F) = 300/600 = 1/2 = .5 = 50\%$

The probability that a student preferred soccer given that the student was a girl is 50%. So 50% of the females prefer soccer. This is a conditional probability because it is only based on females, not on all of the students.

b) Are being female and preferring soccer independent events? Justify your answer. No. Being female and preferring soccer are dependent events because there are females that prefer soccer. The probability of being female and preferring soccer is not 0, it is 300 out of 1000 or 30%.

Algebra 2 Made Easy – Common Core Edition

Algebra 2 Made Easy Handbook

Common Core Standards Edition

Correlation

of

Standards

CORRELATIONS TO COMMON CORE STATE STANDARDS

Correlations